Essentials of Fish Biology

Essentials of Fish Biology

Edited by
Pierce Stubbs

Larsen & Keller
www.larsen-keller.com

Essentials of Fish Biology
Edited by Pierce Stubbs
ISBN: 978-1-63549-760-1 (Hardback)

☰ Larsen & Keller

Published by Larsen and Keller Education,
5 Penn Plaza,
19th Floor,
New York, NY 10001, USA

Cataloging-in-Publication Data

Essentials of fish biology / edited by Pierce Stubbs.
    p. cm.
Includes bibliographical references and index.
ISBN 978-1-63549-760-1
1. Fishes. 2. Ichthyology. I. Stubbs, Pierce.
QL615 .E87 2018
597--dc23

For more information regarding Larsen and Keller Education and its products, please visit the publisher's website www.larsen-keller.com

Table of Contents

Preface

Fish biology refers to the study of the group of animals that have gills and are aquatic craniates. It includes animals like lampreys, bony fish, hagfish, cartilaginous fish, etc. As fishes are an essential food product for animals as well as human beings, it becomes very important to understand their biology. Therefore, fish biology includes the examination of their digestion, circulation, respiration, excretion, scales, etc. This book provides comprehensive insights into the field of fish biology. It picks up individual branches and explains their need and contribution in the context of the growth of this discipline. Those with an interest in fish biology would find this textbook helpful.

A foreword of all Chapters of the book is provided below:

Chapter 1 - The scientific study of fish is known as fish biology. Acanthodii, osteichthyes and agnatha are some of the types of fishes discussed in the section. This is an introductory chapter which will introduce briefly all the significant aspects of fish biology; **Chapter 2** - Predatory fish is a type of fish that feeds upon other fish or animals. Some of these fishes are muskie, salmon, pike, walleye and perch. The fishes preyed upon by other fishes and marine animals are known as forge fishes. They mainly include menhaden, hilsa, halfbeaks and silversides. The different types of fishes discussed in the chapter help the readers in developing a better understanding of fishes; **Chapter 3** - Fish anatomy studies the form and structure of fishes which include the body, skeleton, external organs like gills, skin and scales and internal organs like stomach, kidneys, liver and heart of the fish. The chapter on fish anatomy offers an insightful focus, keeping in mind the complex subject matter; **Chapter 4** - Fish locomotion is the movement of a fish. The different forms of locomotion are anguilliform, sub-carangiform, carangiform and rapid swimming. The aspects elucidated in this chapter are of vital importance, and provide a better understanding of fish locomotion; **Chapter 5** - The method of reproduction in fishes varies. The five modes of reproduction in fishes are ovuliparity, oviparity, ovoviviparity, hemotrophic and histotrophic viviparity. This chapter has been carefully written to provide an easy understanding of the varied types of fish reproduction.

I would like to thank the entire editorial team who made sincere efforts for this book and my family who supported me in my efforts of working on this book. I take this opportunity to thank all those who have been a guiding force throughout my life.

Editor

Understanding Fish Biology

The scientific study of fish is known as fish biology. Acanthodii, osteichthyes and agnatha are some of the types of fishes discussed in the section. This is an introductory chapter which will introduce briefly all the significant aspects of fish biology.

Fish

A fish is any member of a group of animals that consist of all gill-bearing aquatic craniate animals that lack limbs with digits. They form a sister group to the tunicates, together forming the olfactores. Included in this definition are the living hagfish, lampreys, and cartilaginous and bony fish as well as various extinct related groups. Tetrapods emerged within lobe-finned fishes, so cladistically they are fish as well. However, traditionally fish are rendered obsolete or paraphyletic by excluding the tetrapods (i.e., the amphibians, reptiles, birds and mammals which all descended from within the same ancestry). Because in this manner the term "fish" is defined negatively as a paraphyletic group, it is not considered a formal taxonomic grouping in systematic biology. The traditional term pisces (also ichthyes) is considered a typological, but not a phylogenetic classification.

The earliest organisms that can be classified as fish were soft-bodied chordates that first appeared during the Cambrian period. Although they lacked a true spine, they possessed notochords which allowed them to be more agile than their invertebrate counterparts. Fish would continue to evolve through the Paleozoic era, diversifying into a wide variety of forms. Many fish of the Paleozoic developed external armor that protected them from predators. The first fish with jaws appeared in the Silurian period, after which many (such as sharks) became formidable marine predators rather than just the prey of arthropods.

Most fish are ectothermic ("cold-blooded"), allowing their body temperatures to vary as ambient temperatures change, though some of the large active swimmers like white shark and tuna can hold a higher core temperature. Fish are abundant in most bodies of water. They can be found in nearly all aquatic environments, from high mountain streams (e.g., char and gudgeon) to the abyssal and even hadal depths of the deepest oceans (e.g., gulpers and anglerfish). With 33,100 described species, fish exhibit greater species diversity than any other group of vertebrates.

Fish are an important resource for humans worldwide, especially as food. Commercial and subsistence fishers hunt fish in wild fisheries or farm them in ponds or in cages in the ocean. They are also caught by recreational fishers, kept as pets, raised by fishkeepers, and exhibited in public aquaria. Fish have had a role in culture through the ages, serving as deities, religious symbols, and as the subjects of art, books and movies.

Evolution

Dunkleosteus was a gigantic, 10-metre (33 ft) long prehistoric fish of class Placodermi.

Fish do not represent a monophyletic group, and therefore the "evolution of fish" is not studied as a single event.

Early fish from the fossil record are represented by a group of small, jawless, armored fish known as ostracoderms. Jawless fish lineages are mostly extinct. An extant clade, the lampreys may approximate ancient pre-jawed fish. The first jaws are found in Placodermi fossils. The diversity of jawed vertebrates may indicate the evolutionary advantage of a jawed mouth. It is unclear if the advantage of a hinged jaw is greater biting force, improved respiration, or a combination of factors.

Fish may have evolved from a creature similar to a coral-like sea squirt, whose larvae resemble primitive fish in important ways. The first ancestors of fish may have kept the larval form into adulthood (as some sea squirts do today), although perhaps the reverse is the case.

Taxonomy

Fish are a paraphyletic group: that is, any clade containing all fish also contains the tetrapods, which are not fish. For this reason, groups such as the "Class Pisces" seen in older reference works are no longer used in formal classifications.

Traditional classification divide fish into three extant classes, and with extinct forms sometimes classified within the tree, sometimes as their own classes:

- Class Agnatha (jawless fish)
    - Subclass Cyclostomata (hagfish and lampreys)
    - Subclass Ostracodermi (armoured jawless fish) †
- Class Chondrichthyes (cartilaginous fish)
    - Subclass Elasmobranchii (sharks and rays)
    - Subclass Holocephali (chimaeras and extinct relatives)
- Class Placodermi (armoured fish) †
- Class Acanthodii ("spiny sharks", sometimes classified under bony fishes)†

Leedsichthys (left), of the subclass Actinopterygii, is the largest known fish,
with estimates in 2005 putting its maximum size at 16 metres (52 ft)

- Class Osteichthyes (bony fish)

  - Subclass Actinopterygii (ray finned fishes)

  - Subclass Sarcopterygii (fleshy finned fishes, ancestors of tetrapods)

The above scheme is the one most commonly encountered in non-specialist and general works. Many of the above groups are paraphyletic, in that they have given rise to successive groups: Agnathans are ancestral to Chondrichthyes, who again have given rise to Acanthodiians, the ancestors of Osteichthyes. With the arrival of phylogenetic nomenclature, the fishes has been split up into a more detailed scheme, with the following major groups:

- Class Myxini (hagfish)

- Class Pteraspidomorphi † (early jawless fish)

- Class Thelodonti †

- Class Anaspida †

- Class Petromyzontida or Hyperoartia

  - Petromyzontidae (lampreys)

- Class Conodonta (conodonts) †

- Class Cephalaspidomorphi † (early jawless fish)

  - (unranked) Galeaspida †

  - (unranked) Pituriaspida †

  - (unranked) Osteostraci †

- Infraphylum Gnathostomata (jawed vertebrates)

  - Class Placodermi † (armoured fish)

- o Class Chondrichthyes (cartilaginous fish)

- o Class Acanthodii † (spiny sharks)

- o Superclass Osteichthyes (bony fish)

  - Class Actinopterygii (ray-finned fish)

  - Subclass Chondrostei

  - Order Acipenseriformes (sturgeons and paddlefishes)

  - Order Polypteriformes (reedfishes and bichirs).

  - Subclass Neopterygii

  - Infraclass Holostei (gars and bowfins)

  - Infraclass Teleostei (many orders of common fish)

  - Class Sarcopterygii (lobe-finned fish)

  - Subclass Actinistia (coelacanths)

  - Subclass Dipnoi (lungfish)

† – indicates extinct taxon

Some palaeontologists contend that because Conodonta are chordates, they are primitive fish. For a fuller treatment of this taxonomy.

The position of hagfish in the phylum Chordata is not settled. Phylogenetic research in 1998 and 1999 supported the idea that the hagfish and the lampreys form a natural group, the Cyclostomata, that is a sister group of the Gnathostomata.

The various fish groups account for more than half of vertebrate species. There are almost 28,000 known extant species, of which almost 27,000 are bony fish, with 970 sharks, rays, and chimeras and about 108 hagfish and lampreys. A third of these species fall within the nine largest families; from largest to smallest, these families are Cyprinidae, Gobiidae, Cichlidae, Characidae, Loricariidae, Balitoridae, Serranidae, Labridae, and Scorpaenidae. About 64 families are monotypic, containing only one species. The final total of extant species may grow to exceed 32,500.

Diversity

Examples of the Major Classes of Fish

The term "fish" most precisely describes any non-tetrapod craniate (i.e. an animal with a skull and in most cases a backbone) that has gills throughout life and whose limbs, if any, are in the shape of fins. Unlike groupings such as birds or mammals, fish are not a single clade but a paraphyletic collection of taxa, including hagfishes, lampreys, sharks and rays, ray-finned fish, coelacanths, and lungfish. Indeed, lungfish and coelacanths are closer relatives of tetrapods (such as mammals, birds, amphibians, etc.) than of other fish such as ray-finned fish or sharks, so the last common ancestor of all fish is also an ancestor to tetrapods. As paraphyletic groups are no longer recognised in modern systematic biology, the use of the term "fish" as a biological group must be avoided.

Actinopterygii
(Brown trout)

Sarcopterygii
(Coelacanth)

Fish come in many shapes and sizes. This is a sea dragon, a close relative of the seahorse.
Their leaf-like appendages enable them to blend in with floating seaweed.

Many types of aquatic animals commonly referred to as "fish" are not fish in the sense given above; examples include shellfish, cuttlefish, starfish, crayfish and jellyfish. In earlier times, even biologists did not make a distinction – sixteenth century natural historians classified also seals, whales, amphibians, crocodiles, even hippopotamuses, as well as a host of aquatic invertebrates, as fish. However, according to the definition above, all mammals, including cetaceans like whales and dolphins, are not fish. In some contexts, especially in aquaculture, the true fish are referred to as finfish (or fin fish) to distinguish them from these other animals.

A typical fish is ectothermic, has a streamlined body for rapid swimming, extracts oxygen from water using gills or uses an accessory breathing organ to breathe atmospheric oxygen, has two sets of paired fins, usually one or two (rarely three) dorsal fins, an anal fin, and a tail fin, has jaws, has skin that is usually covered with scales, and lays eggs.

Each criterion has exceptions. Tuna, swordfish, and some species of sharks show some warm-blooded adaptations—they can heat their bodies significantly above ambient water temperature. Streamlining and swimming performance varies from fish such as tuna, salmon, and jacks that can cover 10–20 body-lengths per second to species such as eels and rays that swim no more than 0.5 body-lengths per second. Many groups of freshwater fish extract oxygen from the air as well as from the water using a variety of different structures. Lungfish have paired lungs similar to those of tetrapods, gouramis have a structure called the labyrinth organ that performs a similar function, while many catfish, such as *Corydoras* extract oxygen via the intestine or stomach. Body shape and the arrangement of the fins is highly variable, covering such seemingly un-fishlike forms as seahorses, pufferfish, anglerfish, and gulpers. Similarly, the surface of the skin may be naked (as in moray eels), or covered with scales of a variety of different types usually defined as placoid (typical of sharks and rays), cosmoid (fossil lungfish and coelacanths), ganoid (various fossil fish but also living gars and bichirs), cycloid, and ctenoid (these last two are found on most bony fish). There are even fish that live mostly on land or lay their eggs on land near water. Mudskippers feed and interact with one another on mudflats and go underwater to hide in their burrows. A single, an undescribed species of *Phreatobius*, has been called a true "land fish" as this worm-like catfish strictly lives among waterlogged leaf litter. Many species live in underground lakes, underground rivers or aquifers and are popularly known as cavefish.

Fish range in size from the huge 16-metre (52 ft) whale shark to the tiny 8-millimetre (0.3 in) stout infantfish.

Fish species diversity is roughly divided equally between marine (oceanic) and freshwater ecosystems. Coral reefs in the Indo-Pacific constitute the center of diversity for marine fishes, whereas continental freshwater fishes are most diverse in large river basins of tropical rainforests, especially the Amazon, Congo, and Mekong basins. More than 5,600 fish species inhabit Neotropical freshwaters alone, such that Neotropical fishes represent about 10% of all vertebrate species on the Earth. Exceptionally rich sites in the Amazon basin, such as Cantão State Park, can contain more freshwater fish species than occur in all of Europe.

Anatomy and Physiology

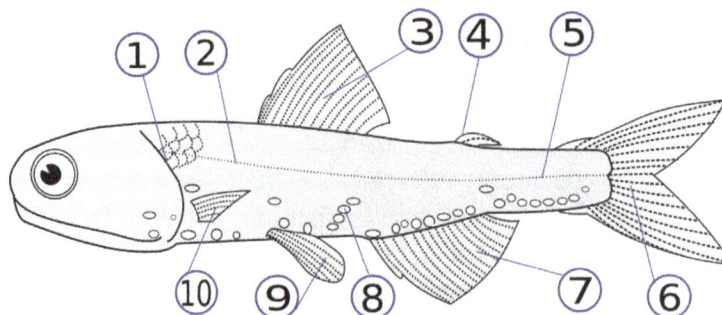

The anatomy of *Lampanyctodes hectoris*
(1) – operculum (gill cover), (2) – lateral line, (3) – dorsal fin, (4) – fat fin, (5) – caudal peduncle,
(6) – caudal fin, (7) – anal fin, (8) – photophores, (9) – pelvic fins (paired), (10) – pectoral fins (paired)

Respiration

Gills

Most fish exchange gases using gills on either side of the pharynx. Gills consist of threadlike structures called filaments. Each filament contains a capillary network that provides a large surface area for exchanging oxygen and carbon dioxide. Fish exchange gases by pulling oxygen-rich water through their mouths and pumping it over their gills. In some fish, capillary blood flows in the opposite direction to the water, causing countercurrent exchange. The gills push the oxygen-poor water out through openings in the sides of the pharynx. Some fish, like sharks and lampreys, possess multiple gill openings. However, bony fish have a single gill opening on each side. This opening is hidden beneath a protective bony cover called an operculum.

Juvenile bichirs have external gills, a very primitive feature that they share with larval amphibians.

Air Breathing

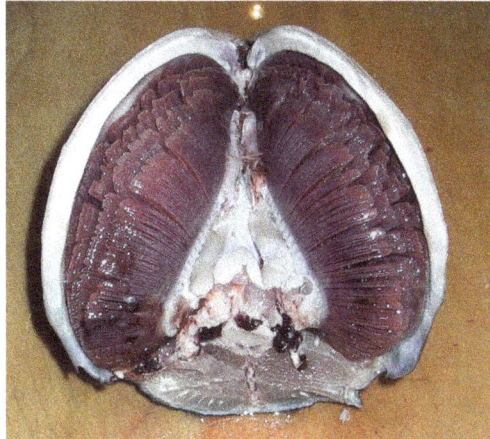

Tuna gills inside the head. The fish head is oriented snout-downwards, with the view looking towards the mouth.

Fish from multiple groups can live out of the water for extended periods. Amphibious fish such as the mudskipper can live and move about on land for up to several days, or live in stagnant or otherwise oxygen depleted water. Many such fish can breathe air via a variety of mechanisms. The skin of anguillid eels may absorb oxygen directly. The buccal cavity of the electric eel may breathe air. Catfish of the families Loricariidae, Callichthyidae, and Scoloplacidae absorb air through their digestive tracts. Lungfish, with the exception of the Australian lungfish, and bichirs have paired lungs similar to those of tetrapods and must surface to gulp fresh air through the mouth and pass spent air out through the gills. Gar and bowfin have a vascularized swim bladder that functions in the same way. Loaches, trahiras, and many catfish breathe by passing air through the gut. Mudskippers breathe by absorbing oxygen across the skin (similar to frogs). A number of fish have evolved so-called accessory breathing organs that extract oxygen from the air. Labyrinth fish (such as gouramis and bettas) have a labyrinth organ above the gills that performs this function. A few other fish have structures resembling labyrinth organs in form and function, most notably snakeheads, pikeheads, and the Clariidae catfish family.

Breathing air is primarily of use to fish that inhabit shallow, seasonally variable waters where the water's oxygen concentration may seasonally decline. Fish dependent solely on dissolved oxygen, such as perch and cichlids, quickly suffocate, while air-breathers survive for much longer, in some cases in water that is little more than wet mud. At the most extreme, some air-breathing fish are able to survive in damp burrows for weeks without water, entering a state of aestivation (summertime hibernation) until water returns.

Air breathing fish can be divided into obligate air breathers and facultative air breathers. Obligate air breathers, such as the African lungfish, *must* breathe air periodically or they suffocate. Facultative air breathers, such as the catfish *Hypostomus plecostomus*, only breathe air if they need to and will otherwise rely on their gills for oxygen. Most air breathing fish are facultative air breathers that avoid the energetic cost of rising to the surface and the fitness cost of exposure to surface predators.

Circulation

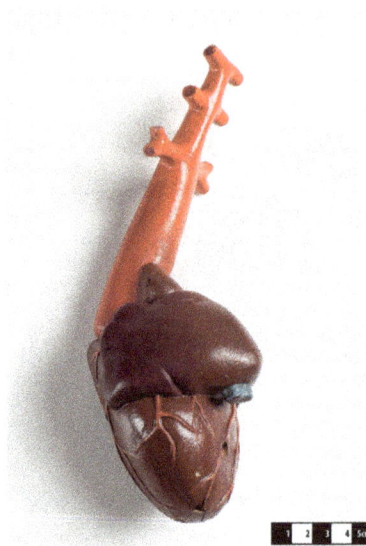

Didactic model of a fish heart.

Fish have a closed-loop circulatory system. The heart pumps the blood in a single loop throughout the body. In most fish, the heart consists of four parts, including two chambers and an entrance and exit. The first part is the sinus venosus, a thin-walled sac that collects blood from the fish's veins before allowing it to flow to the second part, the atrium, which is a large muscular chamber. The atrium serves as a one-way antechamber, sends blood to the third part, ventricle. The ventricle is another thick-walled, muscular chamber and it pumps the blood, first to the fourth part, bulbus arteriosus, a large tube, and then out of the heart. The bulbus arteriosus connects to the aorta, through which blood flows to the gills for oxygenation.

Digestion

Jaws allow fish to eat a wide variety of food, including plants and other organisms. Fish ingest food through the mouth and break it down in the esophagus. In the stomach, food is further digested and, in many fish, processed in finger-shaped pouches called pyloric caeca, which secrete digestive enzymes and absorb nutrients. Organs such as the liver and pancreas add enzymes and various chemicals as the food moves through the digestive tract. The intestine completes the process of digestion and nutrient absorption.

Excretion

As with many aquatic animals, most fish release their nitrogenous wastes as ammonia. Some of the wastes diffuse through the gills. Blood wastes are filtered by the kidneys.

Saltwater fish tend to lose water because of osmosis. Their kidneys return water to the body. The reverse happens in freshwater fish: they tend to gain water osmotically. Their kidneys produce dilute urine for excretion. Some fish have specially adapted kidneys that vary in function, allowing them to move from freshwater to saltwater.

Scales

The scales of fish originate from the mesoderm (skin); they may be similar in structure to teeth.

Sensory and Nervous System

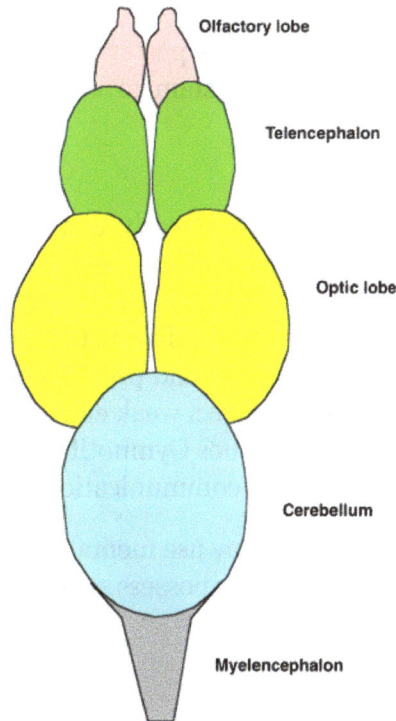

Olfactory lobe

Telencephalon

Optic lobe

Cerebellum

Myelencephalon

Dorsal view of the brain of the rainbow trout

Central Nervous System

Fish typically have quite small brains relative to body size compared with other vertebrates, typically one-fifteenth the brain mass of a similarly sized bird or mammal. However, some fish have relatively large brains, most notably mormyrids and sharks, which have brains about as massive relative to body weight as birds and marsupials.

Fish brains are divided into several regions. At the front are the olfactory lobes, a pair of structures that receive and process signals from the nostrils via the two olfactory nerves. The olfactory lobes are very large in fish that hunt primarily by smell, such as hagfish, sharks, and catfish. Behind the olfactory lobes is the two-lobed telencephalon, the structural equivalent to the cerebrum in higher vertebrates. In fish the telencephalon is concerned mostly with olfaction. Together these structures form the forebrain.

Connecting the forebrain to the midbrain is the diencephalon (in the diagram, this structure is below the optic lobes and consequently not visible). The diencephalon performs functions associated with hormones and homeostasis. The pineal body lies just above the diencephalon. This structure detects light, maintains circadian rhythms, and controls color changes.

The midbrain or mesencephalon contains the two optic lobes. These are very large in species that hunt by sight, such as rainbow trout and cichlids.

The hindbrain or metencephalon is particularly involved in swimming and balance. The cerebellum is a single-lobed structure that is typically the biggest part of the brain. Hagfish and lampreys have relatively small cerebellae, while the mormyrid cerebellum is massive and apparently involved in their electrical sense.

The brain stem or myelencephalon is the brain's posterior. As well as controlling some muscles and body organs, in bony fish at least, the brain stem governs respiration and osmoregulation.

Sense Organs

Most fish possess highly developed sense organs. Nearly all daylight fish have color vision that is at least as good as a human's. Many fish also have chemoreceptors that are responsible for extraordinary senses of taste and smell. Although they have ears, many fish may not hear very well. Most fish have sensitive receptors that form the lateral line system, which detects gentle currents and vibrations, and senses the motion of nearby fish and prey. Some fish, such as catfish and sharks, have the Ampullae of Lorenzini, organs that detect weak electric currents on the order of millivolt. Other fish, like the South American electric fishes Gymnotiformes, can produce weak electric currents, which they use in navigation and social communication.

Fish orient themselves using landmarks and may use mental maps based on multiple landmarks or symbols. Fish behavior in mazes reveals that they possess spatial memory and visual discrimination.

Vision

Vision is an important sensory system for most species of fish. Fish eyes are similar to those of terrestrial vertebrates like birds and mammals, but have a more spherical lens. Their retinas generally have both rods and cones (for scotopic and photopic vision), and most species have colour vision. Some fish can see ultraviolet and some can see polarized light. Amongst jawless fish, the lamprey has well-developed eyes, while the hagfish has only primitive eyespots. Fish vision shows adaptation to their visual environment, for example deep sea fishes have eyes suited to the dark environment.

Hearing

Hearing is an important sensory system for most species of fish. Fish sense sound using their lateral lines and their ears.

Capacity for Pain

Experiments done by William Tavolga provide evidence that fish have pain and fear responses. For instance, in Tavolga's experiments, toadfish grunted when electrically shocked and over time they came to grunt at the mere sight of an electrode.

In 2003, Scottish scientists at the University of Edinburgh and the Roslin Institute concluded that rainbow trout exhibit behaviors often associated with pain in other animals. Bee venom and acetic acid injected into the lips resulted in fish rocking their bodies and rubbing their lips along the sides and floors of their tanks, which the researchers concluded were attempts to relieve pain, similar to what mammals would do. Neurons fired in a pattern resembling human neuronal patterns.

Professor James D. Rose of the University of Wyoming claimed the study was flawed since it did not provide proof that fish possess "conscious awareness, particularly a kind of awareness that is meaningfully like ours". Rose argues that since fish brains are so different from human brains, fish are probably not conscious in the manner humans are, so that reactions similar to human reactions to pain instead have other causes. Rose had published a study a year earlier arguing that fish cannot feel pain because their brains lack a neocortex. However, animal behaviorist Temple Grandin argues that fish could still have consciousness without a neocortex because "different species can use different brain structures and systems to handle the same functions."

Animal welfare advocates raise concerns about the possible suffering of fish caused by angling. Some countries, such as Germany have banned specific types of fishing, and the British RSPCA now formally prosecutes individuals who are cruel to fish.

Muscular System

Swim bladder of a rudd (*Scardinius erythrophthalmus*)

A great white shark off Isla Guadalupe

Most fish move by alternately contracting paired sets of muscles on either side of the backbone. These contractions form S-shaped curves that move down the body. As each curve reaches the back fin, backward force is applied to the water, and in conjunction with the fins, moves the fish forward. The fish's fins function like an airplane's flaps. Fins also increase the tail's surface area, increasing speed. The streamlined body of the fish decreases the amount of friction from the water. Since body tissue is denser than water, fish must compensate for the difference or they will sink. Many bony fish have an internal organ called a swim bladder that adjusts their buoyancy through manipulation of gases.

Homeothermy

Although most fish are exclusively ectothermic, there are exceptions.

Certain species of fish maintain elevated body temperatures. Endothermic teleosts (bony fish) are all in the suborder Scombroidei and include the billfishes, tunas, and one species of "primitive" mackerel (*Gasterochisma melampus*). All sharks in the family Lamnidae – shortfin mako, long fin mako, white, porbeagle, and salmon shark – are endothermic, and evidence suggests the trait exists in family Alopiidae (thresher sharks). The degree of endothermy varies from the billfish, which warm only their eyes and brain, to bluefin tuna and porbeagle sharks who maintain body temperatures elevated in excess of 20 °C above ambient water temperatures. Endothermy, though metabolically costly, is thought to provide advantages such as increased muscle strength, higher rates of central nervous system processing, and higher rates of digestion.

Reproductive System

Organs: 1. Liver, 2. Gas bladder, 3. Roe, 4. Pyloric caeca, 5. Stomach, 6. Intestine

Fish reproductive organs include testicles and ovaries. In most species, gonads are paired organs of similar size, which can be partially or totally fused. There may also be a range of secondary organs that increase reproductive fitness.

In terms of spermatogonia distribution, the structure of teleosts testes has two types: in the most common, spermatogonia occur all along the seminiferous tubules, while in atherinomorph fish they are confined to the distal portion of these structures. Fish can present cystic or semi-cystic spermatogenesis in relation to the release phase of germ cells in cysts to the seminiferous tubules lumen.

Fish ovaries may be of three types: gymnovarian, secondary gymnovarian or cystovarian. In the first type, the oocytes are released directly into the coelomic cavity and then enter the ostium, then through the oviduct and are eliminated. Secondary gymnovarian ovaries shed ova into the coelom from which they go directly into the oviduct. In the third type, the oocytes are conveyed to the exterior through the oviduct. Gymnovaries are the primitive condition found in lungfish, sturgeon, and bowfin. Cystovaries characterize most teleosts, where the ovary lumen has continuity with the oviduct. Secondary gymnovaries are found in salmonids and a few other teleosts.

Oogonia development in teleosts fish varies according to the group, and the determination of oogenesis dynamics allows the understanding of maturation and fertilization processes. Changes in the nucleus, ooplasm, and the surrounding layers characterize the oocyte maturation process.

Postovulatory follicles are structures formed after oocyte release; they do not have endocrine function, present a wide irregular lumen, and are rapidly reabsorbed in a process involving the apoptosis of follicular cells. A degenerative process called follicular atresia reabsorbs vitellogenic oocytes not spawned. This process can also occur, but less frequently, in oocytes in other development stages.

Some fish, like the California sheephead, are hermaphrodites, having both testes and ovaries either at different phases in their life cycle or, as in hamlets, have them simultaneously.

Over 97% of all known fish are oviparous, that is, the eggs develop outside the mother's body. Examples of oviparous fish include salmon, goldfish, cichlids, tuna, and eels. In the majority of these species, fertilisation takes place outside the mother's body, with the male and female fish shedding their gametes into the surrounding water. However, a few oviparous fish practice internal fertilization, with the male using some sort of intromittent organ to deliver sperm into the genital opening of the female, most notably the oviparous sharks, such as the horn shark, and oviparous rays, such as skates. In these cases, the male is equipped with a pair of modified pelvic fins known as claspers.

Marine fish can produce high numbers of eggs which are often released into the open water column. The eggs have an average diameter of 1 millimetre (0.039 in).

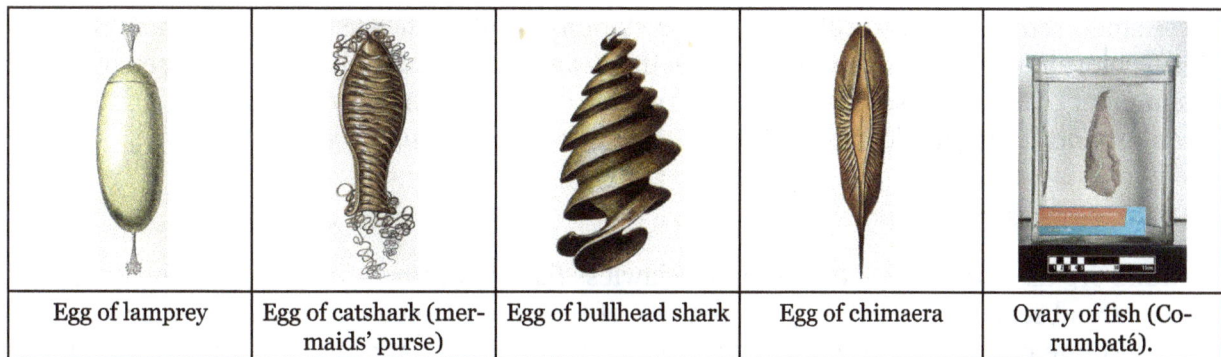

Egg of lamprey	Egg of catshark (mermaids' purse)	Egg of bullhead shark	Egg of chimaera	Ovary of fish (Corumbatá).

The newly hatched young of oviparous fish are called larvae. They are usually poorly formed, carry a large yolk sac (for nourishment), and are very different in appearance from juvenile and adult specimens. The larval period in oviparous fish is relatively short (usually only several weeks), and larvae rapidly grow and change appearance and structure (a process termed metamorphosis) to become juveniles. During this transition larvae must switch from their yolk sac to feeding on zooplankton prey, a process which depends on typically inadequate zooplankton density, starving many larvae.

In ovoviviparous fish the eggs develop inside the mother's body after internal fertilization but receive little or no nourishment directly from the mother, depending instead on the yolk. Each embryo develops in its own egg. Familiar examples of ovoviviparous fish include guppies, angel sharks, and coelacanths.

Some species of fish are viviparous. In such species the mother retains the eggs and nourishes the embryos. Typically, viviparous fish have a structure analogous to the placenta seen in mammals connecting the mother's blood supply with that of the embryo. Examples of viviparous fish include the surf-perches, splitfins, and lemon shark. Some viviparous fish exhibit oophagy, in which the developing embryos eat other eggs produced by the mother. This has been observed primarily among sharks, such as the shortfin mako and porbeagle, but is known for a few bony fish as well, such as the halfbeak *Nomorhamphus ebrardtii*. Intrauterine cannibalism is an even more unusual mode of vivipary, in which the largest embryos eat weaker and smaller siblings. This behavior is also most commonly found among sharks, such as the grey nurse shark, but has also been reported for *Nomorhamphus ebrardtii*.

Aquarists commonly refer to ovoviviparous and viviparous fish as livebearers.

Diseases

Like other animals, fish suffer from diseases and parasites. To prevent disease they have a variety of defenses. *Non-specific* defenses include the skin and scales, as well as the mucus layer secreted by the epidermis that traps and inhibits the growth of microorganisms. If pathogens breach these defenses, fish can develop an inflammatory response that increases blood flow to the infected region and delivers white blood cells that attempt to destroy pathogens. Specific defenses respond to particular pathogens recognised by the fish's body, i.e., an immune response. In recent years, vaccines have become widely used in aquaculture and also with ornamental fish, for example furunculosis vaccines in farmed salmon and koi herpes virus in koi.

Some species use cleaner fish to remove external parasites. The best known of these are the Bluestreak cleaner wrasses of the genus *Labroides* found on coral reefs in the Indian and Pacific oceans. These small fish maintain so-called "cleaning stations" where other fish congregate and perform specific movements to attract the attention of the cleaners. Cleaning behaviors have been observed in a number of fish groups, including an interesting case between two cichlids of the same genus, *Etroplus maculatus*, the cleaner, and the much larger *Etroplus suratensis*.

Immune System

Immune organs vary by type of fish. In the jawless fish (lampreys and hagfish), true lymphoid organs are absent. These fish rely on regions of lymphoid tissue within other organs to produce immune cells. For example, erythrocytes, macrophages and plasma cells are produced in the anterior kidney (or pronephros) and some areas of the gut (where granulocytes mature.) They resemble primitive bone marrow in hagfish. Cartilaginous fish (sharks and rays) have a more advanced immune system. They have three specialized organs that are unique to Chondrichthyes; the epigonal organs (lymphoid tissue similar to mammalian bone) that surround the gonads, the Leydig's organ within the walls of their esophagus, and a spiral valve in their intestine. These organs house typical immune cells (granulocytes, lymphocytes and plasma cells). They also possess an identifiable thymus and a well-developed spleen (their most important immune organ) where various lymphocytes, plasma cells and macrophages develop and are stored. Chondrostean fish (sturgeons, paddlefish, and bichirs) possess a major site for the production of granulocytes within a mass that is associated with the meninges (membranes surrounding the central nervous system.) Their heart is frequently covered with tissue that contains lymphocytes, reticular cells and a small number of macrophages. The chondrostean kidney is an important hemopoietic organ; where erythrocytes, granulocytes, lymphocytes and macrophages develop.

Like chondrostean fish, the major immune tissues of bony fish (or teleostei) include the kidney (especially the anterior kidney), which houses many different immune cells. In addition, teleost fish possess a thymus, spleen and scattered immune areas within mucosal tissues (e.g. in the skin, gills, gut and gonads). Much like the mammalian immune system, teleost erythrocytes, neutrophils and granulocytes are believed to reside in the spleen whereas lymphocytes are the major cell type found in the thymus. In 2006, a lymphatic system similar to that in mammals was described in one species of teleost fish, the zebrafish. Although not confirmed as yet, this system presumably will be where naive (unstimulated) T cells accumulate while waiting to encounter an antigen.

B and T lymphocytes bearing immunoglobulins and T cell receptors, respectively, are found in all jawed fishes. Indeed, the adaptive immune system as a whole evolved in an ancestor of all jawed vertebrate.

Conservation

The 2006 IUCN Red List names 1,173 fish species that are threatened with extinction. Included are species such as Atlantic cod, Devil's Hole pupfish, coelacanths, and great white sharks. Because fish live underwater they are more difficult to study than terrestrial animals and plants, and information about fish populations is often lacking. However, freshwater fish seem particularly threatened because they often live in relatively small water bodies. For example, the Devil's Hole pupfish occupies only a single 3 by 6 metres (10 by 20 ft) pool.

Overfishing

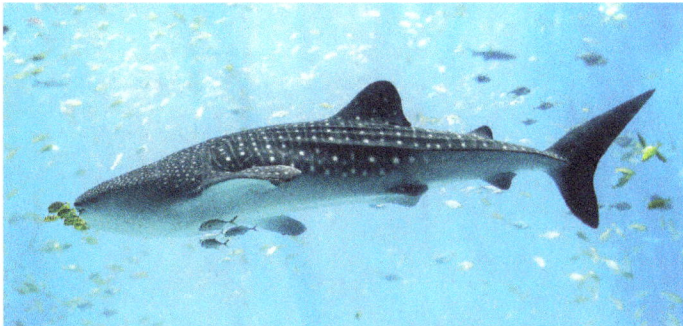

A Whale shark, the world's largest fish, is classified as Vulnerable.

Overfishing is a major threat to edible fish such as cod and tuna. Overfishing eventually causes population (known as stock) collapse because the survivors cannot produce enough young to replace those removed. Such commercial extinction does not mean that the species is extinct, merely that it can no longer sustain a fishery.

One well-studied example of fishery collapse is the Pacific sardine *Sadinops sagax caerulues* fishery off the California coast. From a 1937 peak of 790,000 long tons (800,000 t) the catch steadily declined to only 24,000 long tons (24,000 t) in 1968, after which the fishery was no longer economically viable.

The main tension between fisheries science and the fishing industry is that the two groups have different views on the resiliency of fisheries to intensive fishing. In places such as Scotland, Newfoundland, and Alaska the fishing industry is a major employer, so governments are predisposed to support it. On the other hand, scientists and conservationists push for stringent protection, warning that many stocks could be wiped out within fifty years.

Habitat Destruction

A key stress on both freshwater and marine ecosystems is habitat degradation including water pollution, the building of dams, removal of water for use by humans, and the introduction of exotic species. An example of a fish that has become endangered because of habitat change is the pallid sturgeon, a North American freshwater fish that lives in rivers damaged by human activity.

Exotic Species

Introduction of non-native species has occurred in many habitats. One of the best studied examples is the introduction of Nile perch into Lake Victoria in the 1960s. Nile perch gradually exterminated the lake's 500 endemic cichlid species. Some of them survive now in captive breeding programmes, but others are probably extinct. Carp, snakeheads, tilapia, European perch, brown trout, rainbow trout, and sea lampreys are other examples of fish that have caused problems by being introduced into alien environments.

Importance to Humans

Economic Importance

These fish-farming ponds were created as a cooperative project in a rural village.

Throughout history, humans have utilized fish as a food source. Historically and today, most fish protein has come by means of catching wild fish. However, aquaculture, or fish farming, which has been practiced since about 3,500 BCE. in China, is becoming increasingly important in many nations. Overall, about one-sixth of the world's protein is estimated to be provided by fish. That proportion is considerably elevated in some developing nations and regions heavily dependent on the sea. In a similar manner, fish have been tied to trade.

Catching fish for the purpose of food or sport is known as fishing, while the organized effort by humans to catch fish is called a fishery. Fisheries are a huge global business and provide income for millions of people. The annual yield from all fisheries worldwide is about 154 million tons, with popular species including herring, cod, anchovy, tuna, flounder, and salmon. However, the term fishery is broadly applied, and includes more organisms than just fish, such as mollusks and crustaceans, which are often called "fish" when used as food.

Recreation

Fish have been recognized as a source of beauty for almost as long as used for food, appearing in cave art, being raised as ornamental fish in ponds, and displayed in aquariums in homes, offices, or public settings.

Recreational fishing is fishing for pleasure or competition; it can be contrasted with commercial fishing, which is fishing for profit. The most common form of recreational fishing is done with a rod, reel, line, hooks and any one of a wide range of baits. Angling is a method of fishing, specifically the practice of catching fish by means of an "angle" (hook). Anglers must select the right hook, cast accurately, and retrieve at the right speed while considering water and weather conditions, species, fish response, time of the day, and other factors.

Culture

Avatar of Vishnu as a Matsya

Fish feature prominently in art and literature, in movies such as *Finding Nemo* and books such as *The Old Man and the Sea*. Large fish, particularly sharks, have frequently been the subject of horror movies and thrillers, most notably the novel *Jaws*, which spawned a series of films of the same name that in turn inspired similar films or parodies such as *Shark Tale* and *Snakehead Terror*. Piranhas are shown in a similar light to sharks in films such as *Piranha*; however, contrary to popular belief, the red-bellied piranha is actually a generally timid scavenger species that is unlikely to harm humans. In the Book of Jonah a "great fish" swallowed Jonah the Prophet. Legends of half-human, half-fish mermaids have featured in folklore, including the stories of Hans Christian Andersen.

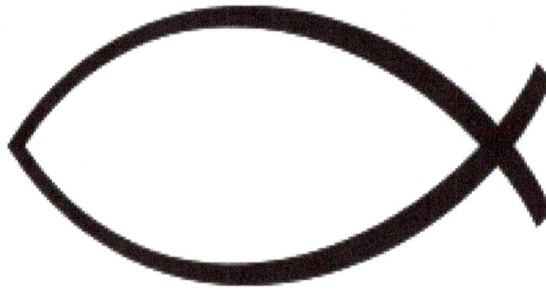

The ichthus is a Christian symbol of a fish signifying that the person who uses it is a Christian.

Fish themes have symbolic significance in many religions. The fish is used often as a symbol by Christians to represent Jesus, or Christianity in general; the gospels also refer to "fishers of men" and feeding the multitude. In the dhamma of Buddhism the fish symbolize happiness as they have complete freedom of movement in the water. Often drawn in the form of carp which are regarded in the Orient as sacred on account of their elegant beauty, size and life-span. Among the deities said to take the form of a fish are Ika-Roa of the Polynesians, Dagon of various ancient Semitic peoples, the shark-gods of Hawai'i and Matsya of the Hindus.

The astrological symbol Pisces is based on a constellation of the same name, but there is also a second fish constellation in the night sky, Piscis Austrinus.

Terminology

Fish or Fishes

Though often used interchangeably, in biology these words have different meanings. *Fish* is used as a singular noun, or as a plural to describe multiple individuals from a single species. *Fishes* is used to describe different species or species groups. Thus a pond that contained a single species might be said to contain 120 fish. But if the pond contained a total of 120 fish from three different species, it would be said to contain three fishes. The distinction is similar to that between people and peoples.

True Fish and Finfish

- In biology, the term *fish* is most strictly used to describe any animal with a backbone that has gills throughout life and has limbs, if any, in the shape of fins. Many types of aquatic animals with common names ending in "fish" are not fish in this sense; examples include shellfish, cuttlefish, starfish, crayfish and jellyfish. In earlier times, even biologists did not make a distinction—sixteenth century natural historians classified also seals, whales, amphibians, crocodiles, even hippopotamuses, as well as a host of aquatic invertebrates, as fish.

- In fisheries, the term *fish* is used as a collective term, and includes mollusks, crustaceans and any aquatic animal which is harvested.

- The strict biological definition of a fish, above, is sometimes called a *true fish*. True fish are also referred to as *finfish* or *fin fish* to distinguish them from other aquatic life harvested in fisheries or aquaculture.

Shoal or School

These goldband fusiliers are schooling because their swimming is synchronised

A random assemblage of fish merely using some localised resource such as food or nesting sites is known simply as an *aggregation*. When fish come together in an interactive, social grouping, then they may be forming either a *shoal* or a *school* depending on the degree of organisation. A *shoal* is a loosely organised group where each fish swims and forages independently but is attracted to other members of the group and adjusts its behaviour, such as swimming speed, so that it remains close to the other members of the group. *Schools* of fish are much more tightly organised, synchronising their swimming so that all fish move at the same speed and in the same direction. Shoaling and schooling behaviour is believed to provide a variety of advantages.

Examples:

- Cichlids congregating at lekking sites form an *aggregation*.

- Many minnows and characins form *shoals*.

- Anchovies, herrings and silversides are classic examples of *schooling* fish.

While the words "school" and "shoal" have different meanings within biology, the distinctions are often ignored by non-specialists who treat the words as synonyms. Thus speakers of British English commonly use "shoal" to describe any grouping of fish, and speakers of American English commonly use "school" just as loosely.

Acanthodii

Acanthodii or acanthodians (sometimes called spiny sharks) is a paraphyletic class of extinct teleostome fish, sharing features with both bony fish and cartilaginous fish. In form they resembled sharks, but their epidermis was covered with tiny rhomboid platelets like the scales of holosteans (gars, bowfins). They represent several independent phylogenetic branch of fishes leading to the still-extant Chondrichthyes.

The popular name "spiny sharks" is a partial misnomer for these early jawed fishes. The name was coined because they were superficially shark-shaped, with a streamlined body, paired fins, and a strongly upturned tail; stout, largely immovable bony spines supporting all the fins except the tail - hence, "spiny sharks"; in retrospect, however, their close relation to modern cartilaginous fish can lead them to be considered "stem-sharks". Fossilized spines and scales are often all that remains of these fishes in ancient sedimentary rocks. Although not sharks or cartilaginous fish, acanthodians did, in fact, have a cartilaginous skeleton, but their fins had a wide, bony base and were reinforced on their anterior margin with a dentine spine. The earliest acanthodians were marine, but during the Devonian, freshwater species became predominant.

There are three orders recognized: Climatiiformes, Ischnacanthiformes and Acanthodiformes. Climatiiforma had shoulder armor and many small sharp spines, Ischnacanthiforma with teeth fused to the jaw, and the Acanthodiforma were filter feeders, with no teeth in the jaw, but long gill rakers. Overall, the acanthodians' jaws are presumed to have evolved from the first gill arch of some ancestral jawless fishes that had a gill skeleton made of pieces of jointed cartilage.

Characteristics

Mesacanthus, Parexus, and *Ischnacanthus* of Early Devonian Great Britain

Diplacanthus longispinus impression at the Museum für Naturkunde, Berlin

The scales of Acanthodii have distinctive ornamentation peculiar to each order. Because of this, the scales are often used in determining relative age of sedimentary rock. The scales are tiny, with a bulbous base, a neck, and a flat or slightly curved diamond-shaped crown.

Despite being called "spiny sharks," acanthodians predate sharks. Scales that have been tentatively identified as belonging to acanthodians, or "shark-like fishes" have been found in various Ordovician strata, though, they are ambiguous, and may actually belong to jawless fishes such as thelodonts. The earliest unequivocal acanthodian fossils date from the beginning of the Silurian Period, some 50 million years before the first sharks appeared. Later, the acanthodians colonized fresh waters, and throve in the rivers and lakes during the Devonian and in the coal swamps of Carboniferous. By this time bony fishes were already showing their potential to dominate the waters of the world, and their competition proved too much for the spiny sharks, which died out in Permian times (approximately 250 Million years ago).

Many palaeontologists originally considered the acanthodians close to the ancestors of the bony fishes. Although their interior skeletons were made of cartilage, a bonelike material had developed in the skins of these fishes, in the form of closely fitting scales. Some scales were greatly enlarged and formed a bony covering on top of the head and over the lower shoulder girdle. Others developed a bony flap over the gill openings analogous to the operculum in later bony fishes. However, most of these characteristics are considered homologous characteristics derived from common placoderm ancestors, and present also in basal cartilaginous fish.

Taxonomy and Phylogeny

In a study of early jawed vertebrate relationships, Davis *et al.* (2012) found acanthodians to be split among the two major clades Osteichthyes (bony fish) and Chondrichthyes (cartilaginous fish). The well-known acanthodian *Acanthodes* was placed within Osteichthyes, despite the presence of many chondrichthyan characteristics in its braincase. However, a newly described Silurian placoderm, *Entelognathus*, which has jaw anatomy shared with bony fish and tetrapods, has led to revisions of this phylogeny: acanthodians are now considered to be a paraphyletic assemblage leading to cartilaginous fish, while bony fish evolved from placoderm ancestors.

Burrow et al. 2016 provides vindication by finding chondrichthyans to be nested among Acanthodii, most closely related to *Doliodus* and *Tamiobatis*. A 2017 study of *Doliodus* morphology points out that it appears to display a mosaic of shark and acanthodian features, making it a transitional fossil and further reinforcing this idea.

Osteichthyes

Osteichthyes, popularly referred to as the bony fish, is a diverse taxonomic group of fish that have skeletons primarily composed of bone tissue, as opposed to cartilage. The vast majority of fish are members of Osteichthyes, which is an extremely diverse and abundant group consisting of 45 orders, and over 435 families and 28,000 species. It is the largest class of vertebrates in existence today. The group Osteichthyes is divided into the ray-finned fish (Actinopterygii) and lobe-finned fish (Sarcopterygii). The oldest known fossils of bony fish are about 420 million years ago, which are also transitional fossils, showing a tooth pattern that is in between the tooth rows of sharks and bony fishes.

Osteichthyes can be compared to Euteleostomi. In paleontology, the terms are synonymous. In ichthyology, the difference is that Euteleostomi presents a cladistic view which includes the terrestrial tetrapods that evolved from lobe-finned fish, whereas on a traditional view, Osteichthyes includes only fishes and is therefore paraphyletic. However, recently published phylogenetic trees treat the Osteichthyes as a clade.

Characteristics

Guiyu oneiros, the earliest known bony fish, lived during the Late Silurian, 419 million years ago).
It has the combination of both ray-finned and lobe-finned features, although analysis
of the totality of its features place it closer to lobe-finned fish.

Bony fish are characterized by a relatively stable pattern of cranial bones, rooted, medial insertion of mandibular muscle in the lower jaw. The head and pectoral girdles are covered with large dermal bones. The eyeball is supported by a sclerotic ring of four small bones, but this characteristic has been lost or modified in many modern species. The labyrinth in the inner ear contains large otoliths. The braincase, or neurocranium, is frequently divided into anterior and posterior sections divided by a fissure.

Early bony fish had simple lungs (a pouch on either side of the esophagus) which helped them breathe in low-oxygen water. In many bony fish these have evolved into swim bladders, which help the body create a neutral balance between sinking and floating. (The lungs of amphibians, reptiles, birds, and mammals were inherited from their bony fish ancestors.) They do not have fin spines, but instead support the fin with lepidotrichia (bone fin rays). They also have an operculum, which helps them breathe without having to swim.

Bony fish have no placoid scales. Mucus glands coat the body. Most have smooth and overlapping ganoid, cycloid or ctenoid scales.

Classification

Traditionally, Osteichthyes is considered a class, recognised on having a swim bladder, only three pairs of gill arches, hidden behind a bony operculum and a predominately bony skeleton. Under this classification systems, the Osteichthyes are paraphyletic with regard to land vertebrates as the common ancestor of all Osteichthyes includes tetrapods amongst its descendants. The largest subclass, the Actinopterygii (ray-finned fish) are monophyletic, but with the inclusion of the smaller sub-class Sarcopterygii, Osteichthyes is paraphyletic.

This has led to an alternative classification, splitting the Osteichthyes into two full classes. Paradoxically, Sarcopterygii is under this scheme monophyletic, as it includes the tetrapods, making it a synonym of the clade Euteleostomi. Most bony fish belong to the ray-finned fish (Actinopterygii).

Actinopterygii ray-finned fish		Actinopterygii, or ray-finned fishes, constitute a class or subclass of the bony fishes. The ray-finned fishes are so called because they possess lepidotrichia or "fin rays", their fins being webs of skin supported by bony or horny spines ("rays"), as opposed to the fleshy, lobed fins that characterize the class Sarcopterygii which also possess lepidotrichia. These actinopterygian fin rays attach directly to the proximal or basal skeletal elements, the radials, which represent the link or connection between these fins and the internal skeleton (e.g., pelvic and pectoral girdles). In terms of numbers, actinopterygians are the dominant class of vertebrates, comprising nearly 99% of the over 30,000 species of fish (Davis, Brian 2010). They are ubiquitous throughout freshwater and marine environments from the deep sea to the highest mountain streams. Extant species can range in size from *Paedocypris*, at 8 mm (0.3 in), to the massive ocean sunfish, at 2,300 kg (5,070 lb), and the long-bodied oarfish, to at least 11 m (36 ft).
Sarcopterygii lobe-finned fish		Sarcopterygii (*fleshy fin*) or lobe-finned fish constitute a clade (traditionally a class or subclass of fish only, i.e. excluding the tetrapods) of the bony fish, though a strict cladistic view includes the terrestrial vertebrates. The living sarcopterygians are the coelacanths, lungfish, and the tetrapods. Early lobe-finned fishes had fleshy, lobed, paired fins, joined to the body by a single bone. Their fins differ from those of all other fish in that each is borne on a fleshy, lobelike, scaly stalk extending from the body. Pectoral and pelvic fins have articulations resembling those of tetrapod limbs. These fins evolved into legs of the first tetrapod land vertebrates, amphibians. They also possess two dorsal fins with separate bases, as opposed to the single dorsal fin of actinopterygians (ray-finned fish). The braincase of sarcoptergygians primitively has a hinge line, but this is lost in tetrapods and lungfish. Many early lobe-finned fishes have a symmetrical tail. All lobe-finned fishes possess teeth covered with true enamel.

Biology

All bony fish possess gills. For the majority this is their sole or main means of respiration. Lungfish and other osteichthyan species are capable of respiration through lungs or vascularized swim bladders. Other species can respire through their skin, intestines, and/or stomach.

Osteichthyes are primitively ectothermic (cold blooded), meaning that their body temperature is dependent on that of the water. But some of the larger marine osteichthyids, such as the opah, swordfish and tuna have independently evolved various levels of endothermy. Bony fish can be any type of heterotroph: numerous species of omnivore, carnivore, herbivore, filter-feeder or detritivore are documented.

Some bony fish are hermaphrodites, and a number of species exhibit parthenogenesis. Fertilization is usually external, but can be internal. Development is usually oviparous (egg-laying) but can be ovoviviparous, or viviparous. Although there is usually no parental care after birth, before birth parents may scatter, hide, guard or brood eggs, with sea horses being notable in that the males undergo a form of "pregnancy", brooding eggs deposited in a ventral pouch by a female.

Examples

The ocean sunfish is the heaviest bony fish in the world, while the longest is the king of herrings, a type of oarfish. Specimens of ocean sunfish have been observed up to 3.3 metres (11 ft) in length and weighing up to 2,303 kilograms (5,077 lb). Other very large bony fish include the Atlantic blue marlin, some specimens of which have been recorded as in excess of 820 kilograms (1,810 lb), the black marlin, some sturgeon species, and the giant and goliath grouper, which both can exceed 300 kilograms (660 lb) in weight. In contrast, the dwarf pygmy goby measures a minute 15 millimetres (0.59 in).

Arapaima gigas is the largest species of freshwater bony fish. The largest bony fish ever was *Leedsichthys*, which dwarfed the beluga sturgeon, ocean sunfish, giant grouper, and all the other giant bony fishes alive today.

Comparison with Cartilaginous Fishes

Cartilaginous fishes can be further divided into sharks, rays and chimaeras. In the table below, the comparison is made between sharks and bony fishes.

Comparison of cartilaginous and bony fishes		
Characteristic	**Sharks (cartilaginous)**	**Bony fishes**
Habitat	Mainly marine	Marine and freshwater
Shape	Usually dorso-ventrally flattened	Usually bilaterally flattened
Exoskeleton	Separate dermal placoid scales	Overlapping dermal cosmoid, ganoid, cycloid or ctenoid scales
Endoskeleton	Cartilaginous	Mostly bony
Caudal fin	Heterocercal	Heterocercal or diphycercal
Pelvic fins	Usually posterior.	Mostly anterior, occasionally posterior.
Intromittent organ	Males use pelvic fins as claspers for transferring sperm to a female	Do not use claspers, though some species use their anal fins as gonopodium for the same purpose
Mouth	Large, crescent shaped on the ventral side of the head	Variable shape and size at the tip or terminal part of the head
Jaw suspension	Hyostylic	Hyostylic and autostylic
Gill openings	Usually five pairs of gill slits which are not protected by an operculum.	Five pairs of gill slits protected by an operculum (a lateral flap of skin).
Type of gills	Larnellibranch with long interbranchial septum	Filiform with reduced interbranchial septum
Spiracles	The first gill slit usually becomes spiracles opening behind the eyes.	No spiracles
Afferent branchial vessels	Five pairs from ventral aorta to gills	Only four pairs
Efferent branchial vessels	Nine pairs	Four pairs
Conus arteriosus	Present in heart	Absent
Cloaca	A true cloaca is present only in cartilaginous fishes and lobe-finned fishes.	In most bony fishes, the cloaca is absent, and the anus, urinary and genital apertures open separately
Stomach	Typically J-shaped	Shape variable. Absent in some.
Intestine	Short with spiral valve in lumen	Long with no spiral valve
Rectal gland	Present	Absent

Liver	Usually has two lobes	Usually has three lobes
Swim bladder	Absent	Usually present
Brain	Has large olfactory lobes and cerebrum with small optic lobes and cerebellum	Has small olfactory lobes and cerebrum and large optic lobes and cerebellum
Restiform bodies	Present in brain	Absent
Ductus endolymphaticus	Opens on top of head	Does not open to exterior
Retina	Lacks cones	Most fish have double cones, a pair of cone cells joined to each other.
Accommodation of eye	Accommodate for near vision by moving the lens closer to the retina	Accommodate for distance vision by moving the lens further from the retina
Ampullae of Lorenzini	Present	Absent
Male genital duct	Connects to the anterior part of the genital kidney	No connection to kidney
Oviducts	Not connected to ovaries	Connected to ovaries
Urinary and genital apertures	United and urinogenital apertures lead into common cloaca	Separate and open independently to exterior
Eggs	A small number of large eggs with plenty of yolk	A large number of small eggs with little yolk
Fertilisation	Internal	Usually external
Development	Ovoviviparous types develop internally. Oviparous types develop externally using egg cases	Normally develop externally without an egg case

Agnatha

Agnatha (Greek, "no jaws") is a superclass of jawless fish in the phylum Chordata, subphylum Vertebrata, consisting of both present (cyclostomes) and extinct (conodonts and ostracoderms) species. The group excludes all vertebrates with jaws, known as gnathostomes.

The agnathans as a whole are paraphyletic, because most extinct agnathans belong to the stem group of gnathostomes. Recent molecular data, both from rRNA and from mtDNA as well as embryological data strongly supports the hypothesis that living agnathans, the cyclostomes, are monophyletic.

The oldest fossil agnathans appeared in the Cambrian, and two groups still survive today: the lampreys and the hagfish, comprising about 120 species in total. Hagfish are considered members of the subphylum Vertebrata, because they secondarily lost vertebrae; before this event was inferred from molecular and developmental data, the group Craniata was created by Linnaeus (and is still sometimes used as a strictly morphological descriptor) to reference hagfish plus vertebrates. In addition to the absence of jaws, modern agnathans are characterised by absence of paired fins; the presence of a notochord both in larvae and adults; and seven or more paired gill pouches. Lampreys have a light sensitive pineal eye (homologous to the pineal gland in mammals). All living and most extinct Agnatha do not have an identifiable stomach or any appendages. Fertilization and development are both external. There is no parental care in the Agnatha class. The Agnatha are ectothermic or cold blooded, with a cartilaginous skeleton, and the heart contains 2 chambers.

While a few scientists still regard the living agnathans as only superficially similar, and argue that many of these similarities are probably shared basal characteristics of ancient vertebrates, recent classifications clearly place hagfish (the Myxini or Hyperotreti), with the lampreys (Hyperoartii) as being more closely related to each other than either is to the jawed fishes.

Metabolism

Agnathans are ectothermic, meaning they do not regulate their own body temperature. Agnathan metabolism is slow in cold water, and therefore they do not have to eat very much. They have no distinct stomach, but rather a long gut, more or less homogenous throughout its length. Lampreys feed on other fish and mammals. They rely on a row of sharp teeth to shred their host. Anticoagulant fluids preventing blood clotting are injected into the host, causing the host to yield more blood. Hagfish are scavengers, eating mostly dead animals. They also use a sharp set of teeth to break down the animal. The fact that Agnathan teeth are unable to move up and down limits their possible food types.

Body Covering

In modern agnathans, the body is covered in skin, with neither dermal or epidermal scales. The skin of hagfish has copious slime glands, the slime constituting their defense mechanism. The slime can sometimes clog up enemy fishes' gills, causing them to die. In direct contrast, many extinct agnathans sported extensive exoskeletons composed of either massive, heavy dermal armour or small mineralized scales.

Appendages

Almost all agnathans, including all extant agnathans, have no paired appendages, although most do have a dorsal or a caudal fin. Some fossil agnathans, such as osteostracans and pituriaspids, did have paired fins, a trait inherited in their jawed descendants.

Reproduction

Fertilization in lampreys is external. Mode of fertilization in hagfishes is not known. Development in both groups probably is external. There is no known parental care. Not much is known about the hagfish reproductive process. It is believed that hagfish only have 30 eggs over a lifetime. Most species are hermaphrodites. There is very little of the larval stage that characterizes the lamprey. Lamprey are only able to reproduce once. After external fertilization, the lamprey's cloacas remain open, allowing a fungus to enter their intestines, killing them. Lampreys reproduce in freshwater riverbeds, working in pairs to build a nest and burying their eggs about an inch beneath the sediment. The resulting hatchlings go through four years of larval development before becoming adults. They also have a certain unusual form of reproduction.

Evolution

Although a minor element of modern marine fauna, agnathans were prominent among the early fish in the early Paleozoic. Two types of Early Cambrian animal apparently having fins, vertebrate musculature, and gills are known from the early Cambrian Maotianshan shales of China: *Haik-*

ouichthys and *Myllokunmingia*. They have been tentatively assigned to Agnatha by Janvier. A third possible agnathid from the same region is *Haikouella*. A possible agnathid that has not been formally described was reported by Simonetti from the Middle Cambrian Burgess Shale of British Columbia.

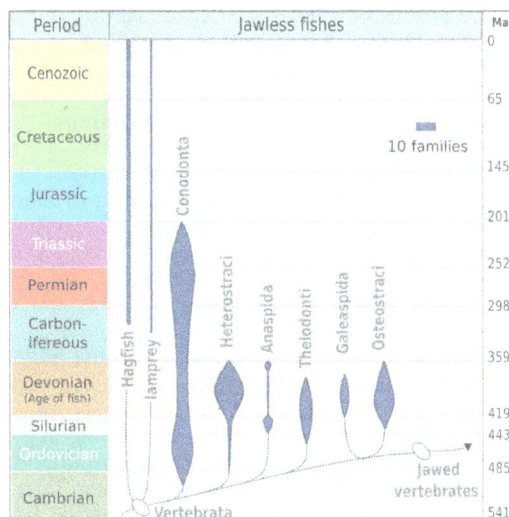

Evolution of jawless fishes. The diagram is based on Michael Benton, 2005.

Many Ordovician, Silurian, and Devonian agnathans were armored with heavy bony-spiky plates. The first armored agnathans—the Ostracoderms, precursors to the bony fish and hence to the tetrapods (including humans)—are known from the middle Ordovician, and by the Late Silurian the agnathans had reached the high point of their evolution. Most of the ostracoderms, such as thelodonts, osteostracans, and galeaspids, were more closely related to the gnathostomes than to the surviving agnathans, known as cyclostomes. Cyclostomes apparently split from other agnathans before the evolution of dentine and bone, which are present in many fossil agnathans, including conodonts. Agnathans declined in the Devonian and never recovered.

References

- Morales, Edwin H. Colbert, Michael (1991). Evolution of the vertebrates : a history of the backboned animals through time (4th ed.). New York: Wiley-Liss. ISBN 978-0-471-85074-8

- "Descubrimiento de fósil de pez óseo en China aporta nuevos conocimientos clave sobre origen de los vertebrados_Spanish.china.org.cn". spanish.china.org.cn. Retrieved 2014-01-25

- Shu, D-G.; Luo, H-L.; Conway Morris, S.; Zhang, X-L.; Hu, S-X.; Chen, L.; Han, J.; Zhu, M.; Li, Y.; Chen, L-Z. (1999). "Lower Cambrian vertebrates from south China". Nature. 402 (6757): 42. doi:10.1038/46965

- Benton, Michael (4 August 2014). Vertebrate Palaeontology. Wiley. p. 281. ISBN 978-1-118-40764-6. Retrieved 22 May 2015

- Schwab, IR; Hart, N (2006). "More than black and white". British Journal of Ophthalmology. 90 (4): 406. PMC 1857009. PMID 16572506. doi:10.1136/bjo.2005.085571

- "Ancient fish face shows roots of modern jaw". Nature. September 25, 2013. Archived from the original on 2013-10-31. Retrieved September 26, 2013

- Stock, David; Whitt GS (1992). "Evidence from 18S ribosomal RNA sequences that lampreys and hagfishes form a natural group". Science. 257 (5071): 787–9. PMID 1496398. doi:10.1126/science.1496398

- Parsons, Alfred Sherwood Romer, Thomas S. (1986). The vertebrate body (6th ed.). Philadelphia: Saunders College Pub. ISBN 978-0-03-910754-3

- Wegner, Nicholas C., Snodgrass, Owen E., Dewar, Heidi, John, Hyde R. Science. "Whole-body endothermy in a mesopelagic fish, the opah, Lampris guttatus". pp. 786–789. Retrieved May 14, 2015

- Betancur-R; et al. (2013). "The Tree of Life and a New Classification of Bony Fishes.". PLOS Currents Tree of Life (Edition 1). doi:10.1371/currents.tol.53ba26640df0ccaee75bb165c8c26288. Archived from the original on 2013-10-13

- Based on: Kotpal R. L. (2010) Modern Text Book Of Zoology Vertebrates Pages 193. Rastogi Publications. ISBN 9788171338917

- Zhao Wen-Jin; Zhu Min (2007). "Diversification and faunal shift of Siluro-Devonian vertebrates of China". Geological Journal. 42 (3–4): 351–369. doi:10.1002/gj.1072

- The early and mid Silurian. See Kazlev, M.A.; White, T (March 6, 2001). "Thelodonti". Palaeos.com. Retrieved October 30, 2007

- Ota, Kinya; et al. (2011). "Identification of vertebra-like elements and their possible differentiation from sclerotomes in the hagfish". Nature Communications. 2 (6): 373. PMC 3157150. PMID 21712821. doi:10.1038/ncomms1355

- Romer, Alfred Sherwood; Parsons, Thomas S. (1977). The Vertebrate Body. Philadelphia, PA: Holt-Saunders International. pp. 396–399. ISBN 0-03-910284-X

- Sansom, Robert S. (2009). "Phylogeny, classification, & character polarity of the Osteostraci (Vertebrata)". Journal of Systematic Palaeontology. 7: 95–115. doi:10.1017/S1477201908002551

- "Warm Blood Makes Opah an Agile Predator". Fisheries Resources Division of the Southwest Fisheries Science Center of the National Oceanic and Atmospheric Administration. May 12, 2015. Retrieved May 15, 2015

- Sarjeant, William Antony S.; L. B. Halstead (1995). Vertebrate fossils and the evolution of scientific concepts: writings in tribute to Beverly Halstead. ISBN 978-2-88124-996-9

Different Types of Fishes

Predatory fish is a type of fish that feeds upon other fish or animals. Some of these fishes are muskie, salmon, pike, walleye and perch. The fishes preyed upon by other fishes and marine animals are known as forge fishes. They mainly include menhaden, hilsa, halfbeaks and silversides. The different types of fishes discussed in the chapter help the readers in developing a better understanding of fishes.

Predatory Fish

A barracuda preying on a smaller fish

Predatory fish are fish that prey upon other fish or animals. Some predatory fish include perch, muskie, pike, walleye and salmon.

Levels of large predatory fish in the global oceans were estimated to be about 10% of their pre-industrial levels by 2003. Large predatory fish are most at risk of extinction; there was a disproportionate level of large predatory fish extinctions during the Cretaceous–Paleogene extinction event 66 million years ago. Creation of marine reserves has been found to restore populations of large predatory fish such as the *Serranidae* — groupers and sea bass.

Predatory fish switch between types of prey in response to variations in their abundance. Such changes in preference are disproportionate and are selected for as evolutionarily efficient. Predatory fish may become a pest if they are introduced into an ecosystem in which they become a new top predator. An example, which has caused much trouble in Maryland and Florida, is the snakehead fish.

Predatory fish such as sharks and tuna form a part of the human diet, but they tend to concentrate significant quantities of mercury in their bodies if they are high in the food chain, especially as apex predators, due to biomagnification.

Predators are an important factor to consider in managing fisheries, and methods for doing so are available and used in some places.

Salmon

Pacific salmon leaping at Willamette Falls, Oregon

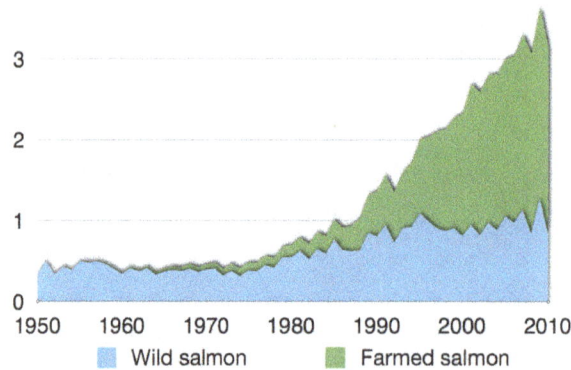

Commercial production of salmon in million tonnes 1950–2010

Salmon is the common name for several species of ray-finned fish in the family Salmonidae. Other fish in the same family include trout, char, grayling and whitefish. Salmon are native to tributaries of the North Atlantic (genus *Salmo*) and Pacific Ocean (genus *Oncorhynchus*). Many species of salmon have been introduced into non-native environments such as the Great Lakes of North America and Patagonia in South America. Salmon are intensively farmed in many parts of the world.

Typically, salmon are anadromous: they are born in fresh water, migrate to the ocean, then return to fresh water to reproduce. However, populations of several species are restricted to fresh water through their lives. Various species of salmon display anadromous life strategies while others display freshwater resident life strategies. Folklore has it that the fish return to the exact spot where they were born to spawn; tracking studies have shown this to be mostly true. A portion of a returning salmon run may stray and spawn in different freshwater systems. The percent of straying depends on the species of salmon. Homing behavior has been shown to depend on olfactory memory.

Species

The term "salmon" comes from the Latin *salmo*, which in turn might have originated from *salire*, meaning "to leap". The nine commercially important species of salmon occur in two genera. The genus *Salmo* contains the Atlantic salmon, found in the north Atlantic, as well as many species commonly named trout. The genus *Oncorhynchus* contains eight species which occur naturally only in the North Pacific. As a group, these are known as Pacific salmon. Chinook salmon have been introduced in New Zealand and Patagonia. Coho, freshwater sockeye, and Atlantic salmon have been established in Patagonia, as well.

Atlantic and Pacific salmon												
Genus	Common name	Scientific name	Maximum length	Common length	Maximum weight	Maximum age	Trophic level	Fish Base	FAO	ITIS	IUCN status
Salmo (Atlantic salmon)	Atlantic salmon	*Salmo salar* Linnaeus, 1758	150 cm	120 cm	46.8 kg	13 years	4.4				LC Least concern
Oncorhynchus (Pacific salmon)	Chinook salmon	*Oncorhynchus tshawytscha* (Walbaum, 1792)	150 cm	70 cm	61.4 kg	9 years	4.4				Not assessed
	Chum salmon	*Oncorhynchus keta* (Walbaum, 1792)	100 cm	58 cm	15.9 kg	7 years	3.5				Not assessed
	Coho salmon	*Oncorhynchus kisutch* (Walbaum, 1792)	108 cm	71 cm	15.2 kg	5 years	4.2				Not assessed
	Masu salmon	*Oncorhynchus masou* (Brevoort, 1856)	79 cm	cm	10.0 kg	3 years	3.6				Not assessed
	Pink salmon	*Oncorhynchus gorbuscha* (Walbaum, 1792)	76 cm	50 cm	6.8 kg	3 years	4.2				Not assessed
	Sockeye salmon	*Oncorhynchus nerka* (Walbaum, 1792)	84 cm	58 cm	7.7 kg	8 years	3.7				LC Least concern

†Both the *Salmo* and *Oncorhynchus* genera also contain a number of species referred to as trout. Within *Salmo*, additional minor taxa have been called salmon in English, i.e. the Adriatic salmon (*Salmo obtusirostris*) and Black Sea salmon (*Salmo labrax*). The steelhead anadromous form of the rainbow trout migrates to sea, but it is not termed "salmon".

Also a number of other species have common names which refer to them as being salmon. Of those listed below, the Danube salmon or *huchen* is a large freshwater salmonid related to the salmon above, but others are marine fishes of the unrelated Perciformes order:

Some other fishes called salmon										
Com-mon name	Scientific name	Maxi-mum length	Com-mon length	Maxi-mum weight	Maxi-mum age	Tro-phic level	Fish Base	FAO	ITIS	IUCN status
Aus-tralian salmon	*Arripis trutta* (Forster, 1801)	89 cm	47 cm	9.4 kg	26 years	4.1				Not as-sessed
Danube salmon	*Hucho hucho* (Linnaeus, 1758)	150 cm	70 cm	52 kg	15 years	4.2				EN Endan-gered
Ha-waiian salmon	*Elagatis bipinnula-ta* (Quoy & Gaimard, 1825)	180 cm	90 cm	46.2 kg	years	3.6				Not as-sessed
Indian salmon	*Eleutherone-ma tetradac-tylum* (Shaw, 1804)	200 cm	50 cm	145 kg	years	4.4				Not as-sessed

Eosalmo driftwoodensis, the oldest known salmon in the fossil record, helps scientists figure how the different species of salmon diverged from a common ancestor. The British Columbia salmon fossil provides evidence that the divergence between Pacific and Atlantic salmon had not yet oc-curred 40 million years ago. Both the fossil record and analysis of mitochondrial DNA suggest the divergence occurred by 10 to 20 million years ago. This independent evidence from DNA analysis and the fossil record rejects the glacial theory of salmon divergence.

Distribution

Atlantic salmon, *Salmo salar*

- Atlantic salmon (*Salmo salar*) reproduce in northern rivers on both coasts of the Atlantic Ocean.

  - Landlocked salmon (*Salmo salar* m. *sebago*) live in a number of lakes in eastern North America and in Northern Europe, for instance in lakes Sebago, Onega, Ladoga, Saimaa, Vänern, and Winnipesaukee. They are not a different species from the Atlantic salmon, but have independently evolved a non-migratory life cycle, which they maintain even when they could access the ocean.

- Chinook salmon (*Oncorhynchus tshawytscha*) are also known in the US as king salmon or blackmouth salmon, and as spring salmon in British Columbia. Chinook are the largest of all Pacific salmon, frequently exceeding 14 kg (30 lb). The name tyee is used in British

Columbia to refer to Chinook over 30 pounds, and in the Columbia River watershed, especially large Chinook were once referred to as June hogs. Chinook salmon are known to range as far north as the Mackenzie River and Kugluktuk in the central Canadian arctic, and as far south as the Central California coast.

• Chum salmon (*Oncorhynchus keta*) are known as dog, keta, or calico salmon in some parts of the US. This species has the widest geographic range of the Pacific species: south to the Sacramento River in California in the eastern Pacific and the island of Kyūshū in the Sea of Japan in the western Pacific; north to the Mackenzie River in Canada in the east and to the Lena River in Siberia in the west.

• Coho salmon (*Oncorhynchus kisutch*) are also known in the US as silver salmon. This species is found throughout the coastal waters of Alaska and British Columbia and as far south as Central California (Monterey Bay). It is also now known to occur, albeit infrequently, in the Mackenzie River.

• Masu salmon or cherry salmon (*Oncorhynchus masou*) are found only in the western Pacific Ocean in Japan, Korea, and Russia. A land-locked subspecies known as the Taiwanese salmon or Formosan salmon (*Oncorhynchus masou formosanus*) is found in central Taiwan's Chi Chia Wan Stream.

• Pink salmon (*Oncorhynchus gorbuscha*), known as humpies in southeast and southwest Alaska, are found from northern California and Korea, throughout the northern Pacific, and from the Mackenzie River in Canada to the Lena River in Siberia, usually in shorter coastal streams. It is the smallest of the Pacific species, with an average weight of 1.6 to 1.8 kg (3.5 to 4.0 lb).

• Sockeye salmon (*Oncorhynchus nerka*) are also known in the US as red salmon. This lake-rearing species is found south as far as the Klamath River in California in the eastern Pacific and northern Hokkaidō island in Japan in the western Pacific and as far north as Bathurst Inlet in the Canadian Arctic in the east and the Anadyr River in Siberia in the west. Although most adult Pacific salmon feed on small fish, shrimp, and squid, sockeye feed on plankton they filter through gill rakers. Kokanee salmon are the land-locked form of sockeye salmon.

• Danube salmon, or huchen (*Hucho hucho*), are the largest permanent freshwater salmonid species.

Life Cycle

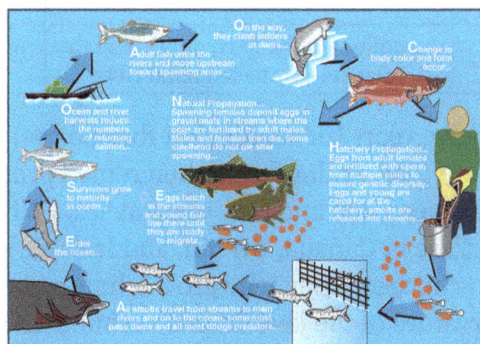

Life cycle of Pacific salmon

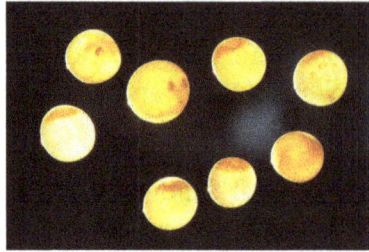

Eggs in different stages of development: In some, only a few cells grow on top of the yolk, in the lower right, the blood vessels surround the yolk, and in the upper left, the black eyes are visible, even the little lens.

Salmon fry hatching — the baby has grown around the remains of the yolk — visible are the arteries spinning around the yolk and little old drops, also the gut, the spine, the main caudal blood vessel, the bladder, and the arcs of the gills

Salmon eggs are laid in freshwater streams typically at high latitudes. The eggs hatch into alevin or sac fry. The fry quickly develop into parr with camouflaging vertical stripes. The parr stay for six months to three years in their natal stream before becoming smolts, which are distinguished by their bright, silvery colour with scales that are easily rubbed off. Only 10% of all salmon eggs are estimated to survive to this stage. The smolt body chemistry changes, allowing them to live in salt water. Smolts spend a portion of their out-migration time in brackish water, where their body chemistry becomes accustomed to osmoregulation in the ocean.

| Male ocean-phase adult sockeye | Juvenile salmon, parr, grow up in the relatively protected natal river | Male spawning-phase adult sockeye |

The salmon spend about one to five years (depending on the species) in the open ocean, where they gradually become sexually mature. The adult salmon then return primarily to their natal streams to spawn. Atlantic salmon spend between one and four years at sea. (When a fish returns after just one year's sea feeding, it is called a grilse in Canada, Britain, and Ireland.) Prior to spawning, depending on the species, salmon undergo changes. They may grow a hump, develop canine-like teeth, or develop a kype (a pronounced curvature of the jaws in male salmon). All change from the silvery blue of a fresh-run fish from the sea to a darker colour. Salmon can make amazing journeys, sometimes moving hundreds of miles upstream against strong currents and rapids to reproduce. Chinook and sockeye salmon from central Idaho, for example, travel over 1,400 km (900 mi) and climb nearly 2,100 m (7,000 ft) from the Pacific Ocean as they return to spawn. Condition tends to deteriorate the longer the fish remain in fresh water, and they then deteriorate further after they spawn, when they are known as kelts. In all species of Pacific salmon, the mature individuals die within a few days or weeks of spawning, a trait known as semelparity. Between 2 and 4% of Atlantic salmon kelts survive to spawn again, all females. However, even in those species of salmon that may survive to spawn more than once (iteroparity), postspawning mortality is quite high (perhaps as high as 40 to 50%.)

To lay her roe, the female salmon uses her tail (caudal fin), to create a low-pressure zone, lifting gravel to be swept downstream, excavating a shallow depression, called a redd. The redd may sometimes contain 5,000 eggs covering 2.8 m² (30 sq ft). The eggs usually range from orange to red. One or more males approach the female in her redd, depositing sperm, or milt, over the roe. The female then covers the eggs by disturbing the gravel at the upstream edge of the depression before moving on to make another redd. The female may make as many as seven redds before her supply of eggs is exhausted.

Each year, the fish experiences a period of rapid growth, often in summer, and one of slower growth, normally in winter. This results in ring formation around an earbone called the otolith, (annuli) analogous to the growth rings visible in a tree trunk. Freshwater growth shows as densely crowded rings, sea growth as widely spaced rings; spawning is marked by significant erosion as body mass is converted into eggs and milt.

Freshwater streams and estuaries provide important habitat for many salmon species. They feed on terrestrial and aquatic insects, amphipods, and other crustaceans while young, and primarily on other fish when older. Eggs are laid in deeper water with larger gravel, and need cool water and good water flow (to supply oxygen) to the developing embryos. Mortality of salmon in the early life stages is usually high due to natural predation and human-induced changes in habitat, such as siltation, high water temperatures, low oxygen concentration, loss of stream cover, and reductions in river flow. Estuaries and their associated wetlands provide vital nursery areas for the salmon prior to their departure to the open ocean. Wetlands not only help buffer the estuary from silt and pollutants, but also provide important feeding and hiding areas.

Salmon not killed by other means show greatly accelerated deterioration (phenoptosis, or "programmed aging") at the end of their lives. Their bodies rapidly deteriorate right after they spawn as a result of the release of massive amounts of corticosteroids.

Ecology

Bears and Salmon

In the Pacific Northwest and Alaska, salmon are keystone species, supporting wildlife such as birds, bears and otters. The bodies of salmon represent a transfer of nutrients from the ocean, rich in nitrogen, sulfur, carbon and phosphorus, to the forest ecosystem.

Grizzly bears function as ecosystem engineers, capturing salmon and carrying them into adjacent wooded areas. There they deposit nutrient-rich urine and feces and partially eaten carcasses. Bears are estimated to leave up to half the salmon they harvest on the forest floor, in densities that can reach 4,000 kilograms per hectare, providing as much as 24% of the total nitrogen available to the riparian woodlands. The foliage of spruce trees up to 500 m (1,600 ft) from a stream where grizzlies fish salmon have been found to contain nitrogen originating from fished salmon.

Bear cub with salmon

Beavers and Salmon

Sockeye salmon jumping over beaver dam

Beavers also function as ecosystem engineers; in the process of clear-cutting and damming, beavers alter their ecosystems extensively. Beaver ponds can provide critical habitat for juvenile salmon. An example of this was seen in the years following 1818 in the Columbia River Basin. In 1818, the British government made an agreement with the U.S. government to allow U.S. citizens access to the Columbia catchment. At the time, the Hudson's Bay Company sent word to trappers to extirpate all furbearers from the area in an effort to make the area less attractive to U.S. fur traders. In response to the elimination of beavers from large parts of the river system, salmon runs plummeted, even in the absence of many of the factors usually associated with the demise of salmon runs. Salmon recruitment can be affected by beavers' dams because dams can:

- Slow the rate at which nutrients are flushed from the system; nutrients provided by adult salmon dying throughout the fall and winter remain available in the spring to newly hatched juveniles

- Provide deeper water pools where young salmon can avoid avian predators

- Increase productivity through photosynthesis and by enhancing the conversion efficiency of the cellulose-powered detritus cycle

- Create low-energy environments where juvenile salmon put the food they ingest into growth rather than into fighting currents

- Increase structural complexity with many physical niches where salmon can avoid predators

Beavers' dams are able to nurture salmon juveniles in estuarine tidal marshes where the salinity is less than 10 ppm. Beavers build small dams of generally less than 60 cm (2 ft) high in channels in the myrtle zone. These dams can be overtopped at high tide and hold water at low tide. This provides refuges for juvenile salmon so they do not have to swim into large channels where they are subject to predation.

Lampreys and Salmon

It has been discovered that rivers which have seen a decline or disappearance of anadromous lampreys also affects the salmon in a negative way. Like salmon, the adults stop feeding and die after spawning, and their decomposing bodies release nutrients into the stream. Their larvae, called ammocoetes, are filter feeders that contributes to the health of the waters. They are a food source for the young salmon, and being fattier and oilier, it is assumed predators prefer them over salmon offspring, taking off some of the predation pressure on smolts.

Parasites

According to Canadian biologist Dorothy Kieser, the myxozoan parasite *Henneguya salminicola* is commonly found in the flesh of salmonids. It has been recorded in the field samples of salmon returning to the Haida Gwaii Islands. The fish responds by walling off the parasitic infection into a number of cysts that contain milky fluid. This fluid is an accumulation of a large number of parasites.

Henneguya salminicola, a myxozoan parasite commonly found in the flesh of salmonids on the West Coast of Canada, in coho salmon

Henneguya and other parasites in the myxosporean group have complex life cycles, where the salmon is one of two hosts. The fish releases the spores after spawning. In the *Henneguya* case, the spores enter a second host, most likely an invertebrate, in the spawning stream. When juvenile salmon migrate to the Pacific Ocean, the second host releases a stage infective to salmon. The parasite is then carried in the salmon until the next spawning cycle. The myxosporean parasite that causes whirling disease in trout has a similar life cycle. However, as opposed to whirling disease, the *Henneguya* infestation does not appear to cause disease in the host salmon — even heavily infected fish tend to return to spawn successfully.

According to Dr. Kieser, a lot of work on *Henneguya salminicola* was done by scientists at the Pacific Biological Station in Nanaimo in the mid-1980s, in particular, an overview report which states, "the fish that have the longest fresh water residence time as juveniles have the most notice-able infections. Hence in order of prevalence coho are most infected followed by sockeye, chinook, chum and pink." As well, the report says, at the time the studies were conducted, stocks from the middle and upper reaches of large river systems in British Columbia such as Fraser, Skeena, Nass and from mainland coastal streams in the southern half of B.C., "are more likely to have a low prev-alence of infection." The report also states, "It should be stressed that *Henneguya*, economically deleterious though it is, is harmless from the view of public health. It is strictly a fish parasite that cannot live in or affect warm blooded animals, including man".

According to Klaus Schallie, Molluscan Shellfish Program Specialist with the Canadian Food In-spection Agency, "*Henneguya salminicola* is found in southern B.C. also and in all species of salm-on. I have previously examined smoked chum salmon sides that were riddled with cysts and some sockeye runs in Barkley Sound (southern B.C., west coast of Vancouver Island) are noted for their high incidence of infestation."

Sea lice, particularly *Lepeophtheirus salmonis* and various *Caligus* species, including *C. clemensi* and *C. rogercresseyi*, can cause deadly infestations of both farm-grown and wild salmon. Sea lice are ectoparasites which feed on mucus, blood, and skin, and migrate and latch onto the skin of wild salmon during free-swimming, planktonic nauplii and copepodid larval stages, which can persist for several days. Large numbers of highly populated, open-net salmon farms can create exception-ally large concentrations of sea lice; when exposed in river estuaries containing large numbers of open-net farms, many young wild salmon are infected, and do not survive as a result. Adult salmon may survive otherwise critical numbers of sea lice, but small, thin-skinned juvenile salmon migrat-ing to sea are highly vulnerable. On the Pacific coast of Canada, the louse-induced mortality of pink salmon in some regions is commonly over 80%.

Effect of Pile Driving

The risk of injury caused by underwater pile driving has been studied by Dr Halvorsen and her co-workers. The study concluded that the fish are at risk of injury if the cumulative sound exposure level exceeds 210 dB relative to $1 \ \mu Pa^2$ s.

Wild Fisheries

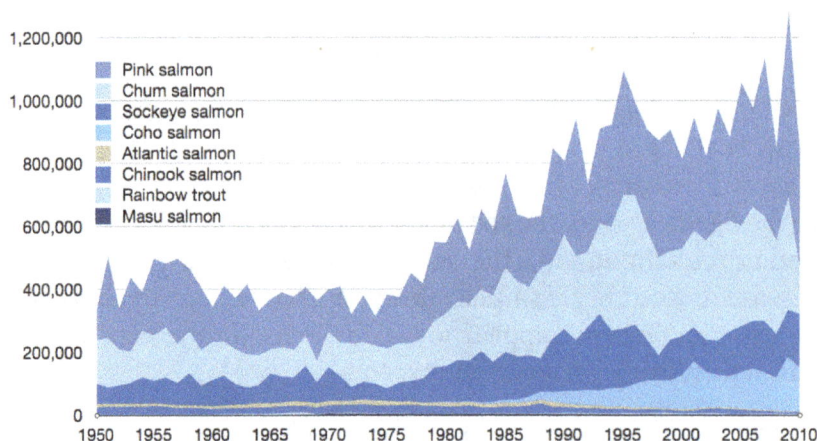

Wild fisheries – commercial capture of all true wild salmon species 1950–2010, as reported by the FAO

Commercial

As can be seen from the production chart at the left, the global capture reported by different countries to the FAO of commercial wild salmon has remained fairly steady since 1990 at about one million tonnes per year. This is in contrast to farmed salmon (below) which has increased in the same period from about 0.6 million tonnes to well over two million tonnes.

Nearly all captured wild salmon are Pacific salmon. The capture of wild Atlantic salmon has always been relatively small, and has declined steadily since 1990. In 2011 only 2,500 tonnes were reported. In contrast about half of all farmed salmon are Atlantic salmon.

Angler and gillie land a salmon, Scotland

Recreational

Recreational salmon fishing can be a technically demanding kind of sport fishing, not necessarily congenial for beginning fishermen. A conflict exists between commercial fishermen and recreational fishermen for the right to salmon stock resources. Commercial fishing in estuaries and coastal areas is often restricted so enough salmon can return to their natal rivers where they can spawn and be available for sport fishing. On parts of the North American west coast sport salmon fishing completely replaces inshore commercial fishing. The commercial value of a salmon can be several times less than the value of the same fish caught by a sport fisherman. This is "a powerful economic argument for allocating stock resources preferentially to sport fishing."

Farmed Salmon

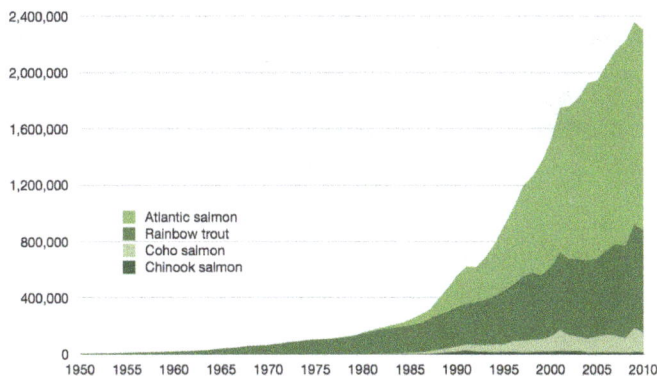

Aquaculture production of all true salmon species 1950–2010, as reported by the FAO

Salmon aquaculture is a major contributor to the world production of farmed finfish, representing about US$10 billion annually. Other commonly cultured fish species include: tilapia, catfish, sea bass, carp and bream. Salmon farming is significant in Chile, Norway, Scotland, Canada and the Faroe Islands; it is the source for most salmon consumed in the United States and Europe. Atlantic salmon are also, in very small volumes, farmed in Russia and the island of Tasmania, Australia.

Salmon are carnivorous. They are fed a meal produced from catching other wild fish and other marine organisms. Salmon farming leads to a high demand for wild forage fish. Salmon require large nutritional intakes of protein, and farmed salmon consume more fish than they generate as a final product. To produce one pound of farmed salmon, products from several pounds of wild fish are fed to them. As the salmon farming industry expands, it requires more wild forage fish for feed, at a time when 75% of the world's monitored fisheries are already near to or have exceeded their maximum sustainable yield. The industrial-scale extraction of wild forage fish for salmon farming affects the survivability of the wild predator fish which rely on them for food.

Work continues on substituting vegetable proteins for animal proteins in the salmon diet. This substitution results in lower levels of the highly valued omega-3 fatty acid content in the farmed product.

Intensive salmon farming uses open-net cages, which have low production costs. It has the drawback of allowing disease and sea lice to spread to local wild salmon stocks.

On a dry weight basis, 2–4 kg of wild-caught fish are needed to produce one kg of salmon.

Artificially incubated chum salmon

Another form of salmon production, which is safer but less controllable, is to raise salmon in hatcheries until they are old enough to become independent. They are released into rivers in an attempt to increase the salmon population. This system is referred to as ranching. It was very common in countries such as Sweden, before the Norwegians developed salmon farming, but is seldom done by private companies. As anyone may catch the salmon when they return to spawn, a company is limited in benefiting financially from their investment.

Rainbow trout farm in an archipelago of Finland

Because of this, the ranching method has mainly been used by various public authorities and non-profit groups, such as the Cook Inlet Aquaculture Association, as a way to increase salmon populations in situations where they have declined due to overharvesting, construction of dams, and habitat destruction or fragmentation. Negative consequences to this sort of population manipulation include genetic "dilution" of the wild stocks. Many jurisdictions are now beginning to discourage supplemental fish planting in favour of harvest controls, and habitat improvement and protection.

A variant method of fish stocking, called ocean ranching, is under development in Alaska. There, the young salmon are released into the ocean far from any wild salmon streams. When it is time for them to spawn, they return to where they were released, where fishermen can catch them.

An alternative method to hatcheries is to use spawning channels. These are artificial streams, usually parallel to an existing stream, with concrete or rip-rap sides and gravel bottoms. Water from the adjacent stream is piped into the top of the channel, sometimes via a header pond, to settle out sediment. Spawning success is often much better in channels than in adjacent streams due to the control of floods, which in some years can wash out the natural redds. Because of the lack of floods, spawning channels must sometimes be cleaned out to remove accumulated sediment. The same floods that destroy natural redds also clean the regular streams. Spawning channels preserve the natural selection of natural streams, as there is no benefit, as in hatcheries, to use prophylactic chemicals to control diseases.

Farm-raised salmon are fed the carotenoids astaxanthin and canthaxanthin to match their flesh colour to wild salmon to improve their marketability.

One proposed alternative to the use of wild-caught fish as feed for the salmon, is the use of soy-based products. This should be better for the local environment of the fish farm, but producing soy beans has a high environmental cost for the producing region.

Another possible alternative is a yeast-based coproduct of bioethanol production, proteinaceous fermentation biomass. Substituting such products for engineered feed can result in equal (sometimes enhanced) growth in fish. With its increasing availability, this would address the problems of rising costs for buying hatchery fish feed.

Yet another attractive alternative is the increased use of seaweed. Seaweed provides essential minerals and vitamins for growing organisms. It offers the advantage of providing natural amounts of dietary fiber and having a lower glycemic load than grain-based fish meal. In the best-case scenario, widespread use of seaweed could yield a future in aquaculture that eliminates the need for land, freshwater, or fertilizer to raise fish.

Management

Spawning sockeye salmon in Becharof Creek, Becharof Wilderness, Alaska

The population of wild salmon declined markedly in recent decades, especially North Atlantic populations, which spawn in the waters of western Europe and eastern Canada, and wild salmon in the Snake and Columbia River systems in northwestern United States.

Salmon population levels are of concern in the Atlantic and in some parts of the Pacific. Alaska fishery stocks are still abundant, and catches have been on the rise in recent decades, after the state initiated limitations in 1972. Some of the most important Alaskan salmon sustainable wild fisheries are located near the Kenai River, Copper River, and in Bristol Bay. Fish farming of Pacific salmon is outlawed in the United States Exclusive Economic Zone, however, there is a substantial network of publicly funded hatcheries, and the State of Alaska's fisheries management system is viewed as a leader in the management of wild fish stocks. In Canada, returning Skeena River wild salmon support commercial, subsistence and recreational fisheries, as well as the area's diverse wildlife on the coast and around communities hundreds of miles inland in the watershed. The status of wild salmon in Washington is mixed. Of 435 wild stocks of salmon and steelhead, only 187 of them were classified as healthy; 113 had an unknown status, one was extinct, 12 were in critical condition and 122 were experiencing depressed populations.

The commercial salmon fisheries in California have been either severely curtailed or closed completely in recent years, due to critically low returns on the Klamath and or Sacramento rivers, causing millions of dollars in losses to commercial fishermen. Both Atlantic and Pacific salmon are popular sportfish.

Salmon populations have been established in all the Great Lakes. Coho stocks were planted by the state of Michigan in the late 1960s to control the growing population of non-native alewife. Now Chinook (king), Atlantic, and coho (silver) salmon are annually stocked in all Great Lakes by most bordering states and provinces. These populations are not self-sustaining and do not provide much in the way of a commercial fishery, but have led to the development of a thriving sport fishery.

As Food

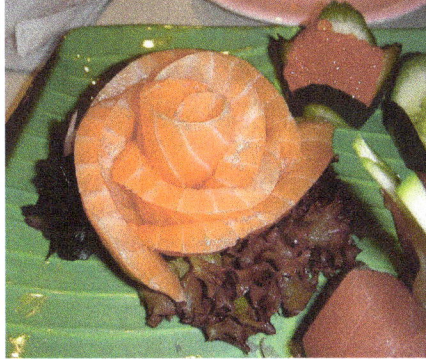
Salmon sashimi

Salmon is a popular food. Classified as an oily fish, salmon is considered to be healthy due to the fish's high protein, high omega-3 fatty acids, and high vitamin D content. Salmon is also a source of cholesterol, with a range of 23–214 mg/100 g depending on the species. According to reports in the journal *Science*, farmed salmon may contain high levels of dioxins. PCB (polychlorinated biphenyl) levels may be up to eight times higher in farmed salmon than in wild salmon, but still well below levels considered dangerous. Nonetheless, according to a 2006 study published in the Journal of the American Medical Association, the benefits of eating even farmed salmon still outweigh any risks imposed by contaminants. Farmed salmon has a high omega 3 fatty acid content comparable to wild salmon. The type of omega-3 present may not be a factor for other important health functions.

Salmon flesh is generally orange to red, although white-fleshed wild salmon with white-black skin colour occurs. The natural colour of salmon results from carotenoid pigments, largely astaxanthin, but also canthaxanthin, in the flesh. Wild salmon get these carotenoids from eating krill and other tiny shellfish.

The vast majority of Atlantic salmon available around the world are farmed (almost 99%), whereas the majority of Pacific salmon are wild-caught (greater than 80%). Canned salmon in the US is usually wild Pacific catch, though some farmed salmon is available in canned form. Smoked salmon is another popular preparation method, and can either be hot or cold smoked. Lox can refer to either cold-smoked salmon or salmon cured in a brine solution (also called gravlax). Traditional canned salmon includes some skin (which is harmless) and bone (which adds calcium). Skinless and boneless canned salmon is also available.

Raw salmon flesh may contain *Anisakis* nematodes, marine parasites that cause anisakiasis. Before the availability of refrigeration, the Japanese did not consume raw salmon. Salmon and salmon roe have only recently come into use in making sashimi (raw fish) and sushi.

To the Indigenous peoples of the Pacific Northwest Coast, salmon is considered a vital part of the diet. Specifically, the indigenous peoples of Haida Gwaii, located near former Queen Charlotte Island in British Columbia, rely on salmon as one of their main sources of food, although many other bands have fished Pacific waters for centuries. Salmon are not only ancient and unique, but it is important because it is expressed in culture, art forms, and ceremonial feasts. Annually, salmon spawn in Haida, feeding on everything on the way upstream and down. Within the Haida nation, salmon is referred to as *"tsiin"*, and is prepared in several ways including smoking, baking, frying, and making soup.

Historically, there has always been enough salmon, as people would not overfish, and only took what they needed. In 2003, a report on First Nation participation in commercial fisheries, including salmon, commissioned by BC's Ministry of Agriculture, Food and Fisheries found that there were 595 First Nation-owned and operated commercial vessels in the province. Of those vessels, First Nations' members owned 564. However, employment within the industry has decreased overall by 50% in the last decade, with 8,142 registered commercial fishermen in 2003. This has affected employment for many fisherman, who rely on salmon as a source of income.

Black bears also rely on salmon as food. The leftovers the bears leave behind are considered important nutrients for the forest, such as the soil, trees, and plants. In this sense, the salmon feed the forest and in return receive clean water and gravel in which to hatch and grow, sheltered from extremes of temperature and water flow in times of high and low rainfall. However, the condition of the salmon in Haida has been affected in recent decades. Due to logging and development, much of the salmon's habitat (i.e.: Ain River) has been destroyed, resulting in the fish being close to endangered. For residents, this has resulted in limits on catches, in turn, has affected families diets, and cultural events such as feasts. Some of the salmon systems in danger include: the Davidon, Naden, Mamim, and Mathers. It is clear that further protection is needed for salmon, such as their habitats, where logging commonly occurs.

History

Seine fishing for salmon – Wenzel Hollar, 1607–1677

The salmon has long been at the heart of the culture and livelihood of coastal dwellers. Many people of the northern Pacific shore had a ceremony to honor the first return of the year. For many centuries, people caught salmon as they swam upriver to spawn. A famous spearfishing site on the Columbia River at Celilo Falls was inundated after great dams were built on the river. The Ainu, of northern Japan, trained dogs to catch salmon as they returned to their breeding grounds *en masse*. Now, salmon are caught in bays and near shore.

The Columbia River salmon population is now less than 3% of what it was when Lewis and Clark arrived at the river. Salmon canneries established by settlers beginning in 1866 had a strong negative impact on the salmon population. In his 1908 State of the Union address, U.S. President Theodore Roosevelt observed that the fisheries were in significant decline:

The salmon fisheries of the Columbia River are now but a fraction of what they were twenty-five years ago, and what they would be now if the United States Government had taken complete charge of them by intervening between Oregon and Washington. During these twenty-five years the fishermen of each State have naturally tried to take all they could get, and the two legislatures have never been able to agree on joint action of any kind adequate in degree for the protection of the fisheries. At the moment the fishing on the Oregon side is practically closed, while there is no limit on the Washington side of any kind, and no one can tell what the courts will decide as to the very statutes under which this action and non-action result. Meanwhile very few salmon reach the spawning grounds, and probably four years hence the fisheries will amount to nothing; and this comes from a struggle between the associated, or gill-net, fishermen on the one hand, and the owners of the fishing wheels up the river.

On the Columbia River the Chief Joseph Dam completed in 1955 completely blocks salmon migration to the upper Columbia River system.

The Fraser River salmon population was affected by the 1914 slide caused by the Canadian Pacific Railway at Hells Gate. The 1917 catch was one quarter of the 1913 catch.

Mythology

Scales on the "Big Fish" or "Salmon of Knowledge" celebrates the return of fish to the River Lagan

The salmon is an important creature in several strands of Celtic mythology and poetry, which often associated them with wisdom and venerability. In Irish mythology, a creature called the Salmon of Knowledge plays key role in the tale *The Boyhood Deeds of Fionn*. In the tale, the Salmon will grant powers of knowledge to whoever eats it, and is sought by poet Finn Eces for seven years. Finally Finn Eces catches the fish and gives it to his young pupil, Fionn mac Cumhaill, to prepare it for him. However, Fionn burns his thumb on the salmon's juices, and he instinctively puts it in his mouth. In so doing, he inadvertently gains the Salmon's wisdom. Elsewhere in Irish mythology, the salmon is also one of the incarnations of both Tuan mac Cairill and Fintan mac Bóchra.

Salmon also feature in Welsh mythology. In the prose tale *Culhwch and Olwen*, the Salmon of Llyn Llyw is the oldest animal in Britain, and the only creature who knows the location of Mabon ap Modron. After speaking to a string of other ancient animals who do not know his whereabouts, King Arthur's men Cai and Bedwyr are led to the Salmon of Llyn Llyw, who lets them ride its back to the walls of Mabon's prison in Gloucester.

In Norse mythology, after Loki tricked the blind god Höðr into killing his brother Baldr, Loki jumped into a river and transformed himself into a salmon to escape punishment from the other gods. When they held out a net to trap him he attempted to leap over it but was caught by Thor who grabbed him by the tail with his hand, and this is why the salmon's tail is tapered.

Salmon are central spiritually and culturally to Native American mythology on the Pacific coast, from the Haida and Coast Salish peoples, to the Nuu-chah-nulth peoples in British Columbia.

Shark

Sharks are a group of elasmobranch fish characterized by a cartilaginous skeleton, five to seven gill slits on the sides of the head, and pectoral fins that are not fused to the head. Modern sharks are classified within the clade Selachimorpha (or Selachii) and are the sister group to the rays. However, the term "shark" has also been used for extinct members of the subclass Elasmobranchii outside the Selachimorpha, such as *Cladoselache* and *Xenacanthus*, as well as other Chondrichthyes such as the holocephalid eugenedontidans. Under this broader definition, the earliest known sharks date back to more than 420 million years ago. Acanthodians are often referred to as "spiny sharks"; though they are not part of Chondrichthyes proper, they are a paraphyletic assemblage leading to cartilaginous fish as a whole.

Since then, sharks have diversified into over 500 species. They range in size from the small dwarf lanternshark (*Etmopterus perryi*), a deep sea species of only 17 centimetres (6.7 in) in length, to the whale shark (*Rhincodon typus*), the largest fish in the world, which reaches approximately 12 metres (40 ft) in length. Sharks are found in all seas and are common to depths of 2,000 metres (6,600 ft). They generally do not live in freshwater although there are a few known exceptions, such as the bull shark and the river shark, which can survive and be found in both seawater and freshwater. Sharks have a covering of dermal denticles that protects their skin from damage and parasites in addition to improving their fluid dynamics. They have numerous sets of replaceable teeth.

Well-known species such as the great white shark, tiger shark, blue shark, mako shark, thresher shark, and the hammerhead shark are apex predators—organisms at the top of their underwater food chain. Many shark populations are threatened by human activities.

Etymology

Until the 16th century, sharks were known to mariners as "sea dogs". This is still evidential in several species termed "dogfish," or the porbeagle.

The etymology of the word "shark" is uncertain, the most likely etymology states that the original sense of the word was that of "predator, one who preys on others" from the Dutch *schurk*, meaning "villain, scoundrel" (cf. *card shark*, *loan shark*, etc.), which was later applied to the fish due to its predatory behaviour.

A now disproven theory is that it derives from the Yucatec Maya word *xok*, pronounced 'shok'. Evidence for this etymology came from the Oxford English Dictionary, which notes *shark* first came into use after Sir John Hawkins' sailors exhibited one in London in 1569 and posted "*sharke*" to refer to the large sharks of the Caribbean Sea. However, the Middle English Dictionary records an isolated occurrence of the word *shark* (referring to a sea fish) in a letter written by Thomas Beckington in 1442, which rules out a New World etymology.

Evolution

A collection of Cretaceous shark teeth

Evidence for the existence of sharks dates from the Ordovician period, 450–420 million years ago, before land vertebrates existed and before many plants had colonized the continents. Only scales have been recovered from the first sharks and not all paleontologists agree that these are from true sharks, suspecting that these scales are actually those of thelodont agnathans. The oldest generally accepted shark scales are from about 420 million years ago, in the Silurian period. The first sharks looked very different from modern sharks. The majority of modern sharks can be traced back to around 100 million years ago. Most fossils are of teeth, often in large numbers. Partial skeletons and even complete fossilized remains have been discovered. Estimates suggest that sharks grow tens of thousands of teeth over a lifetime, which explains the abundant fossils. The teeth consist of easily fossilized calcium phosphate, an apatite. When a shark dies, the decomposing skeleton breaks up, scattering the apatite prisms. Preservation requires rapid burial in bottom sediments.

Among the most ancient and primitive sharks is *Cladoselache*, from about 370 million years ago, which has been found within Paleozoic strata in Ohio, Kentucky, and Tennessee. At that point in Earth's history these rocks made up the soft bottom sediments of a large, shallow ocean, which stretched across much of North America. *Cladoselache* was only about 1 metre (3.3 ft) long with stiff triangular fins and slender jaws. Its teeth had several pointed cusps, which wore down from use. From the small number of teeth found together, it is most likely that *Cladoselache* did not replace its teeth as regularly as modern sharks. Its caudal fins had a similar shape to the great white sharks and the pelagic shortfin and longfin makos. The presence of whole fish arranged tail-first in their stomachs suggest that they were fast swimmers with great agility.

Most fossil sharks from about 300 to 150 million years ago can be assigned to one of two groups. The Xenacanthida was almost exclusive to freshwater environments. By the time this group became extinct about 220 million years ago, they had spread worldwide. The other group, the hybodonts, appeared about 320 million years ago and lived mostly in the oceans, but also in freshwater. The results of a 2014 study of the gill structure of an unusually well-preserved 325 million year old fossil suggested that sharks are not "living fossils", but rather have evolved more extensively than previously thought over the hundreds of millions of years they have been around.

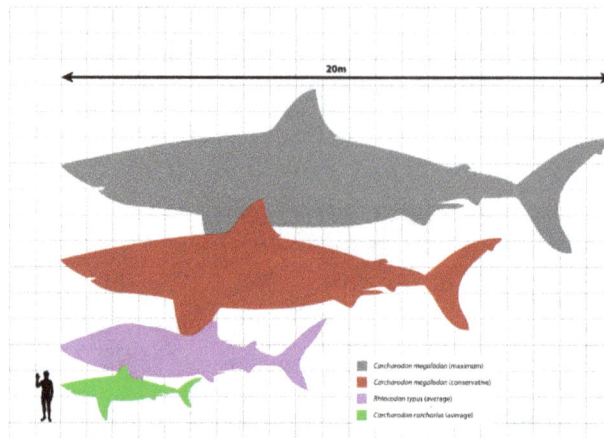

Megalodon (top two, estimated maximum and conservative sizes)
with the whale shark, great white shark, and a human for scale

Modern sharks began to appear about 100 million years ago. Fossil mackerel shark teeth date to the Early Cretaceous. One of the most recently evolved families is the hammerhead shark (family Sphyrnidae), which emerged in the Eocene. The oldest white shark teeth date from 60 to 66 million years ago, around the time of the extinction of the dinosaurs. In early white shark evolution there are at least two lineages: one lineage is of white sharks with coarsely serrated teeth and it probably gave rise to the modern great white shark, and another lineage is of white sharks with finely serrated teeth. These sharks attained gigantic proportions and include the extinct mega-toothed shark, *C. megalodon*. Like most extinct sharks, *C. megalodon* is also primarily known from its fossil teeth and vertebrae. This giant shark reached a total length (TL) of more than 16 metres (52 ft). *C. megalodon* may have approached a maxima of 20.3 metres (67 ft) in total length and 103 metric tons (114 short tons) in mass. Paleontological evidence suggests that this shark was an active predator of large cetaceans.

Taxonomy

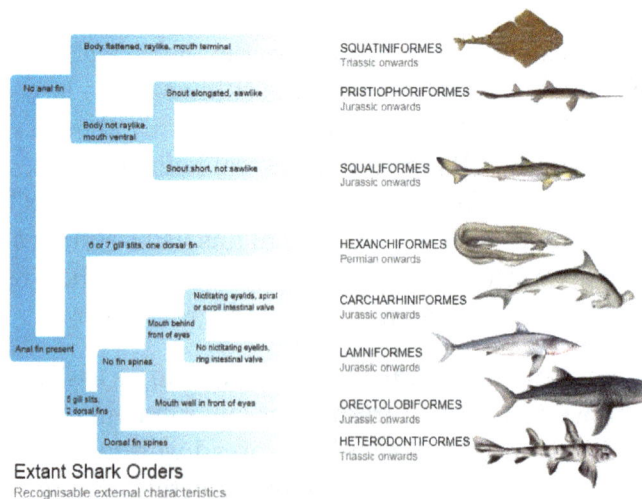

Extant Shark Orders
Recognisable external characteristics

Sharks belong to the superorder Selachimorpha in the subclass Elasmobranchii in the class Chondrichthyes. The Elasmobranchii also include rays and skates; the Chondrichthyes also include Chimaeras. It is currently thought that the sharks form a polyphyletic group: some sharks are more closely related to rays than they are to some other sharks.

The superorder Selachimorpha is divided into Galea (or Galeomorphii), and Squalea (or Squalomorphii). The Galeans are the Heterodontiformes, Orectolobiformes, Lamniformes, and Carcharhiniformes. Lamnoids and Carcharhinoids are usually placed in one clade, but recent studies show the Lamnoids and Orectoloboids are a clade. Some scientists now think that Heterodontoids may be Squalean. The Squaleans are divided into Hexanchoidei and Squalomorpha. The Hexanchoidei includes the Hexanchiformes and Chlamydoselachiformes. The Squalomorpha contains the Squaliformes and the Hypnosqualea. The Hypnosqualea may be invalid. It includes the Squatiniformes, and the Pristorajea, which may also be invalid, but includes the Pristiophoriformes and the Batoidea.

There are more than 470 species of sharks split across thirteen orders, including four orders of sharks that have gone extinct:

- Carcharhiniformes: Commonly known as ground sharks, the order includes the blue, tiger, bull, grey reef, blacktip reef, Caribbean reef, blacktail reef, whitetip reef, and oceanic whitetip sharks (collectively called the requiem sharks) along with the houndsharks, catsharks, and hammerhead sharks. They are distinguished by an elongated snout and a nictitating membrane which protects the eyes during an attack.

- Heterodontiformes: They are generally referred to as the bullhead or horn sharks.

- Hexanchiformes: Examples from this group include the cow sharks and frilled sharks, which somewhat resembles a marine snake.

- Lamniformes: They are commonly known as the mackerel sharks. They include the goblin shark, basking shark, megamouth shark, the thresher sharks, shortfin and longfin mako sharks, and great white shark. They are distinguished by their large jaws and ovoviviparous reproduction. The Lamniformes also include the extinct megalodon, *Carcharodon megalodon*.

- Orectolobiformes: They are commonly referred to as the carpet sharks, including zebra sharks, nurse sharks, wobbegongs, and the whale shark.

- Pristiophoriformes: These are the sawsharks, with an elongated, toothed snout that they use for slashing their prey.

- Squaliformes: This group includes the dogfish sharks and roughsharks.

- Squatiniformes: Also known as angel sharks, they are flattened sharks with a strong resemblance to stingrays and skates.

- † Cladoselachiformes

- † Hybodontiformes

- † Symmoriida

- † Xenacanthida (Xenacantiformes)

Anatomy

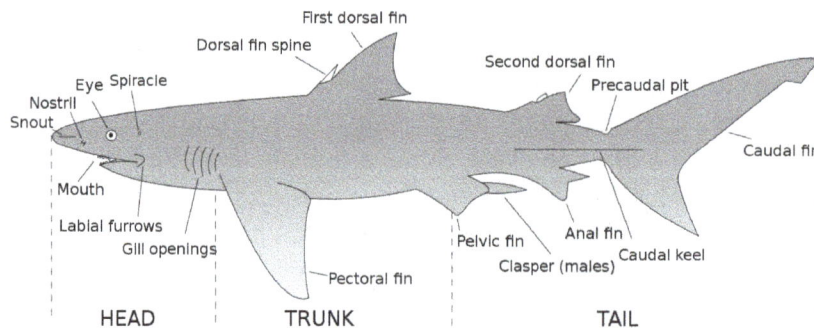

General anatomical features of sharks

Teeth

The teeth of tiger sharks are oblique and serrated to saw through flesh

Shark teeth are embedded in the gums rather than directly affixed to the jaw, and are constantly replaced throughout life. Multiple rows of replacement teeth grow in a groove on the inside of the jaw and steadily move forward in comparison to a conveyor belt; some sharks lose 30,000 or more teeth in their lifetime. The rate of tooth replacement varies from once every 8 to 10 days to several months. In most species, teeth are replaced one at a time as opposed to the simultaneous replacement of an entire row, which is observed in the cookiecutter shark.

Tooth shape depends on the shark's diet: those that feed on mollusks and crustaceans have dense and flattened teeth used for crushing, those that feed on fish have needle-like teeth for gripping, and those that feed on larger prey such as mammals have pointed lower teeth for gripping and triangular upper teeth with serrated edges for cutting. The teeth of plankton-feeders such as the basking shark are small and non-functional.

Skeleton

Shark skeletons are very different from those of bony fish and terrestrial vertebrates. Sharks and other cartilaginous fish (skates and rays) have skeletons made of cartilage and connective tissue.

Cartilage is flexible and durable, yet is about half the normal density of bone. This reduces the skeleton's weight, saving energy. Because sharks do not have rib cages, they can easily be crushed under their own weight on land.

Jaw

Jaws of sharks, like those of rays and skates, are not attached to the cranium. The jaw's surface (in comparison to the shark's vertebrae and gill arches) needs extra support due to its heavy exposure to physical stress and its need for strength. It has a layer of tiny hexagonal plates called "tesserae", which are crystal blocks of calcium salts arranged as a mosaic. This gives these areas much of the same strength found in the bony tissue found in other animals.

Generally sharks have only one layer of tesserae, but the jaws of large specimens, such as the bull shark, tiger shark, and the great white shark, have two to three layers or more, depending on body size. The jaws of a large great white shark may have up to five layers. In the rostrum (snout), the cartilage can be spongy and flexible to absorb the power of impacts.

Fins

Fin skeletons are elongated and supported with soft and unsegmented rays named ceratotrichia, filaments of elastic protein resembling the horny keratin in hair and feathers. Most sharks have eight fins. Sharks can only drift away from objects directly in front of them because their fins do not allow them to move in the tail-first direction.

Dermal Denticles

Unlike bony fish, sharks have a complex dermal corset made of flexible collagenous fibers and arranged as a helical network surrounding their body. This works as an outer skeleton, providing attachment for their swimming muscles and thus saving energy. Their dermal teeth give them hydrodynamic advantages as they reduce turbulence when swimming.

Tails

Tails provide thrust, making speed and acceleration dependent on tail shape. Caudal fin shapes vary considerably between shark species, due to their evolution in separate environments. Sharks possess a heterocercal caudal fin in which the dorsal portion is usually noticeably larger than the ventral portion. This is because the shark's vertebral column extends into that dorsal portion, providing a greater surface area for muscle attachment. This allows more efficient locomotion among these negatively buoyant cartilaginous fish. By contrast, most bony fish possess a homocercal caudal fin.

Tiger sharks have a large upper lobe, which allows for slow cruising and sudden bursts of speed. The tiger shark must be able to twist and turn in the water easily when hunting to support its varied diet, whereas the porbeagle shark, which hunts schooling fish such as mackerel and herring, has a large lower lobe to help it keep pace with its fast-swimming prey. Other tail adaptations help sharks catch prey more directly, such as the thresher shark's usage of its powerful, elongated upper lobe to stun fish and squid.

Physiology

Buoyancy

Unlike bony fish, sharks do not have gas-filled swim bladders for buoyancy. Instead, sharks rely on a large liver filled with oil that contains squalene, and their cartilage, which is about half the normal density of bone. Their liver constitutes up to 30% of their total body mass. The liver's effectiveness is limited, so sharks employ dynamic lift to maintain depth while swimming. Sand tiger sharks store air in their stomachs, using it as a form of swim bladder. Bottom-dwelling sharks, like the nurse shark, have negative buoyancy, allowing them to rest on the ocean floor.

Some sharks, if inverted or stroked on the nose, enter a natural state of tonic immobility. Researchers use this condition to handle sharks safely.

Respiration

Like other fish, sharks extract oxygen from seawater as it passes over their gills. Unlike other fish, shark gill slits are not covered, but lie in a row behind the head. A modified slit called a spiracle lies just behind the eye, which assists the shark with taking in water during respiration and plays a major role in bottom–dwelling sharks. Spiracles are reduced or missing in active pelagic sharks. While the shark is moving, water passes through the mouth and over the gills in a process known as "ram ventilation". While at rest, most sharks pump water over their gills to ensure a constant supply of oxygenated water. A small number of species have lost the ability to pump water through their gills and must swim without rest. These species are *obligate ram ventilators* and would presumably asphyxiate if unable to move. Obligate ram ventilation is also true of some pelagic bony fish species.

The respiration and circulation process begins when deoxygenated blood travels to the shark's two-chambered heart. Here the shark pumps blood to its gills via the ventral aorta artery where it branches into afferent brachial arteries. Reoxygenation takes place in the gills and the reoxygenated blood flows into the efferent brachial arteries, which come together to form the dorsal aorta. The blood flows from the dorsal aorta throughout the body. The deoxygenated blood from the body then flows through the posterior cardinal veins and enters the posterior cardinal sinuses. From there blood enters the heart ventricle and the cycle repeats.

Thermoregulation

Most sharks are "cold-blooded" or, more precisely, poikilothermic, meaning that their internal body temperature matches that of their ambient environment. Members of the family Lamnidae (such as the shortfin mako shark and the great white shark) are homeothermic and maintain a higher body temperature than the surrounding water. In these sharks, a strip of aerobic red muscle located near the center of the body generates the heat, which the body retains via a countercurrent exchange mechanism by a system of blood vessels called the rete mirabile ("miraculous net"). The common thresher shark has a similar mechanism for maintaining an elevated body temperature, which is thought to have evolved independently.

Osmoregulation

In contrast to bony fish, with the exception of the coelacanth, the blood and other tissue of sharks and Chondrichthyes is generally isotonic to their marine environments because of the high concentration of urea (up to 2.5%) and trimethylamine N-oxide (TMAO), allowing them to be in osmotic balance with the seawater. This adaptation prevents most sharks from surviving in freshwater, and they are therefore confined to marine environments. A few exceptions exist, such as the bull shark, which has developed a way to change its kidney function to excrete large amounts of urea. When a shark dies, the urea is broken down to ammonia by bacteria, causing the dead body to gradually smell strongly of ammonia.

Digestion

Digestion can take a long time. The food moves from the mouth to a J-shaped stomach, where it is stored and initial digestion occurs. Unwanted items may never get past the stomach, and instead the shark either vomits or turns its stomachs inside out and ejects unwanted items from its mouth.

One of the biggest differences between the digestive systems of sharks and mammals is that sharks have much shorter intestines. This short length is achieved by the spiral valve with multiple turns within a single short section instead of a long tube-like intestine. The valve provides a long surface area, requiring food to circulate inside the short gut until fully digested, when remaining waste products pass into the cloaca.

Senses

Smell

The shape of the hammerhead shark's head may enhance olfaction by spacing the nostrils further apart.

Sharks have keen olfactory senses, located in the short duct (which is not fused, unlike bony fish) between the anterior and posterior nasal openings, with some species able to detect as little as one part per million of blood in seawater.

Sharks have the ability to determine the direction of a given scent based on the timing of scent detection in each nostril. This is similar to the method mammals use to determine direction of sound.

They are more attracted to the chemicals found in the intestines of many species, and as a result often linger near or in sewage outfalls. Some species, such as nurse sharks, have external barbels that greatly increase their ability to sense prey.

Sight

Eye of a Bigeyed sixgill shark (*Hexanchus nakamurai*)

Shark eyes are similar to the eyes of other vertebrates, including similar lenses, corneas and retinas, though their eyesight is well adapted to the marine environment with the help of a tissue called tapetum lucidum. This tissue is behind the retina and reflects light back to it, thereby increasing visibility in the dark waters. The effectiveness of the tissue varies, with some sharks having stronger nocturnal adaptations. Many sharks can contract and dilate their pupils, like humans, something no teleost fish can do. Sharks have eyelids, but they do not blink because the surrounding water cleans their eyes. To protect their eyes some species have nictitating membranes. This membrane covers the eyes while hunting and when the shark is being attacked. However, some species, including the great white shark (*Carcharodon carcharias*), do not have this membrane, but instead roll their eyes backwards to protect them when striking prey. The importance of sight in shark hunting behavior is debated. Some believe that electro- and chemoreception are more significant, while others point to the nictating membrane as evidence that sight is important. Presumably, the shark would not protect its eyes were they unimportant. The use of sight probably varies with species and water conditions. The shark's field of vision can swap between monocular and stereoscopic at any time. A micro-spectrophotometry study of 17 species of shark found 10 had only rod photoreceptors and no cone cells in their retinas giving them good night vision while making them colorblind. The remaining seven species had in addition to rods a single type of cone photoreceptor sensitive to green and, seeing only in shades of grey and green, are believed to be effectively colorblind. The study indicates that an object's contrast against the background, rather than colour, may be more important for object detection.

Hearing

Although it is hard to test the hearing of sharks, they may have a sharp sense of hearing and can possibly hear prey from many miles away. A small opening on each side of their heads (not the spiracle) leads directly into the inner ear through a thin channel. The lateral line shows a similar arrangement, and is open to the environment via a series of openings called lateral line pores. This is a reminder of the common origin of these two vibration- and sound-detecting organs that are grouped together as the acoustico-lateralis system. In bony fish and tetrapods the external opening into the inner ear has been lost.

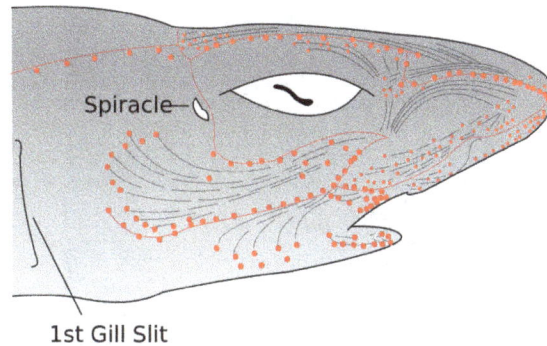

Electromagnetic field receptors (ampullae of Lorenzini) and motion detecting canals in the head of a shark

Electroreception

The ampullae of Lorenzini are the electroreceptor organs. They number in the hundreds to thousands. Sharks use the ampullae of Lorenzini to detect the electromagnetic fields that all living things produce. This helps sharks (particularly the hammerhead shark) find prey. The shark has the greatest electrical sensitivity of any animal. Sharks find prey hidden in sand by detecting the electric fields they produce. Ocean currents moving in the magnetic field of the Earth also generate electric fields that sharks can use for orientation and possibly navigation.

Lateral Line

This system is found in most fish, including sharks. It detects motion or vibrations in water. The shark can sense frequencies in the range of 25 to 50 Hz.

Life history

The claspers of male spotted wobbegong

Shark lifespans vary by species. Most live 20 to 30 years. The spiny dogfish has one of the longest lifespans at more than 100 years. Whale sharks (*Rhincodon typus*) may also live over 100 years. Earlier estimates suggested the Greenland shark (*Somniosus microcephalus*) could reach about 200 years, but a recent study found that a 5.02-metre-long (16.5 ft) specimen was 392 ± 120 years old (i.e., at least 272 years old), making it the longest-lived vertebrate known.

Shark egg

The spiral egg case of a Port Jackson shark

Reproduction

Unlike most bony fish, sharks are K-selected reproducers, meaning that they produce a small number of well-developed young as opposed to a large number of poorly developed young. Fecundity in sharks ranges from 2 to over 100 young per reproductive cycle. Sharks mature slowly relative to many other fish. For example, lemon sharks reach sexual maturity at around age 13–15.

Sexual

Sharks practice internal fertilization. The posterior part of a male shark's pelvic fins are modified into a pair of intromittent organs called claspers, analogous to a mammalian penis, of which one is used to deliver sperm into the female.

Mating has rarely been observed in sharks. The smaller catsharks often mate with the male curling around the female. In less flexible species the two sharks swim parallel to each other while the male inserts a clasper into the female's oviduct. Females in many of the larger species have bite marks that appear to be a result of a male grasping them to maintain position during mating. The bite marks may also come from courtship behavior: the male may bite the female to show his interest. In some species, females have evolved thicker skin to withstand these bites.

Asexual

There have been a number of documented cases in which a female shark who has not been in contact with a male has conceived a sharklet on her own through parthenogenesis. The details of this process are not well understood, but genetic fingerprinting showed that the sharklets had no paternal genetic contribution, ruling out sperm storage. The extent of this behavior in the wild is unknown, as is whether other species have this capability. Mammals are now the only major vertebrate group in which asexual reproduction has not been observed.

Scientists say that asexual reproduction in the wild is rare, and probably a last-ditch effort to reproduce when a mate is not present. Asexual reproduction diminishes genetic diversity, which helps build defenses against threats to the species. Species that rely solely on it risk extinction. Asexual reproduction may have contributed to the blue shark's decline off the Irish coast.

Brooding

Sharks display three ways to bear their young, varying by species, oviparity, viviparity and ovoviviparity.

Ovoviviparity:

Most sharks are ovoviviparous, meaning that the eggs hatch in the oviduct within the mother's body and that the egg's yolk and fluids secreted by glands in the walls of the oviduct nourishes the embryos. The young continue to be nourished by the remnants of the yolk and the oviduct's fluids. As in viviparity, the young are born alive and fully functional. Lamniforme sharks practice *oophagy*, where the first embryos to hatch eat the remaining eggs. Taking this a step further, sand tiger shark sharklets cannibalistically consume neighboring embryos. The survival strategy for ovoviviparous species is to brood the young to a comparatively large size before birth. The whale shark is now classified as ovoviviparous rather than oviparous, because extrauterine eggs are now thought to have been aborted. Most ovoviviparous sharks give birth in sheltered areas, including bays, river mouths and shallow reefs. They choose such areas for protection from predators (mainly other sharks) and the abundance of food. Dogfish have the longest known gestation period of any shark, at 18 to 24 months. Basking sharks and frilled sharks appear to have even longer gestation periods, but accurate data are lacking.

Oviparity

Some species are oviparous like most other fish, laying their eggs in the water. In most oviparous shark species, an egg case with the consistency of leather protects the developing embryo(s). These cases may be corkscrewed into crevices for protection. Once empty, the egg case is known as a *mermaid's purse*, and can wash up on shore. Oviparous sharks include the horn shark, catshark, Port Jackson shark, and swellshark.

Viviparity

Finally some sharks maintain a *placental* link to the developing young, this method is called viviparity. This is more analogous to mammalian gestation than that of other fishes. The young are born alive and fully functional. Hammerheads, the requiem sharks (such as the bull and blue sharks), and smoothhounds are viviparous.

Behavior

The classic view describes a solitary hunter, ranging the oceans in search of food. However, this applies to only a few species. Most live far more social, sedentary, benthic lives, and appear likely to have their own distinct personalities. Even solitary sharks meet for breeding or at rich hunting grounds, which may lead them to cover thousands of miles in a year. Shark migration patterns may be even more complex than in birds, with many sharks covering entire ocean basins.

Sharks can be highly social, remaining in large schools. Sometimes more than 100 scalloped hammerheads congregate around seamounts and islands, e.g., in the Gulf of California. Cross-species social hierarchies exist. For example, oceanic whitetip sharks dominate silky sharks of comparable size during feeding.

When approached too closely some sharks perform a threat display. This usually consists of exaggerated swimming movements, and can vary in intensity according to the threat level.

Speed

In general, sharks swim ("cruise") at an average speed of 8 kilometres per hour (5.0 mph), but when feeding or attacking, the average shark can reach speeds upwards of 19 kilometres per hour (12 mph). The shortfin mako shark, the fastest shark and one of the fastest fish, can burst at speeds up to 50 kilometres per hour (31 mph). The great white shark is also capable of speed bursts. These exceptions may be due to the warm-blooded, or homeothermic, nature of these sharks' physiology. Sharks can travel 70 to 80 km in a day.

Intelligence

Sharks possess brain-to-body mass ratios that are similar to mammals and birds, and have exhibited apparent curiosity and behavior resembling play in the wild.

There is evidence that juvenile lemon sharks can use observational learning in their investigation of novel objects in their environment.

Sleep

All sharks need to keep water flowing over their gills in order for them to breathe, however not all species need to be moving to do this. Those that are able to breathe while not swimming do so by using their spiracles to force water over their gills, thereby allowing them to extract oxygen from the water. It has been recorded that their eyes remain open while in this state and actively follow the movements of divers swimming around them and as such they are not truly asleep.

Species that do need to swim continuously to breathe go through a process known as sleep swimming, in which the shark is essentially unconscious. It is known from experiments conducted on the spiny dogfish that its spinal cord, rather than its brain, coordinates swimming, so spiny dogfish can continue to swim while sleeping, and this also may be the case in larger shark species.

Ecology

Feeding

Most sharks are carnivorous. Basking sharks, whale sharks, and megamouth sharks have independently evolved different strategies for filter feeding plankton: basking sharks practice ram feeding, whale sharks use suction to take in plankton and small fishes, and megamouth sharks make suction feeding more efficient by using the luminescent tissue inside of their mouths to attract prey in the deep ocean. This type of feeding requires gill rakers—long, slender filaments that form a very efficient sieve—analogous to the baleen plates of the great whales. The shark traps the plankton in these filaments and swallows from time to time in huge mouthfuls. Teeth in these species are comparatively small because they are not needed for feeding.

Unlike many other sharks, the great white shark is not actually an apex predator in all of its natural environments, as it is sometimes hunted by orcas

Other highly specialized feeders include cookiecutter sharks, which feed on flesh sliced out of other larger fish and marine mammals. Cookiecutter teeth are enormous compared to the animal's size. The lower teeth are particularly sharp. Although they have never been observed feeding, they are believed to latch onto their prey and use their thick lips to make a seal, twisting their bodies to rip off flesh.

Some seabed–dwelling species are highly effective ambush predators. Angel sharks and wobbegongs use camouflage to lie in wait and suck prey into their mouths. Many benthic sharks feed solely on crustaceans which they crush with their flat molariform teeth.

Other sharks feed on squid or fish, which they swallow whole. The viper dogfish has teeth it can point outwards to strike and capture prey that it then swallows intact. The great white and other large predators either swallow small prey whole or take huge bites out of large animals. Thresher sharks use their long tails to stun shoaling fishes, and sawsharks either stir prey from the seabed or slash at swimming prey with their tooth-studded rostra.

Many sharks, including the whitetip reef shark are cooperative feeders and hunt in packs to herd and capture elusive prey. These social sharks are often migratory, traveling huge distances around ocean basins in large schools. These migrations may be partly necessary to find new food sources.

Range and Habitat

Sharks are found in all seas. They generally do not live in fresh water, with a few exceptions such as the bull shark and the river shark which can swim both in seawater and freshwater. Sharks are common down to depths of 2,000 metres (7,000 ft), and some live even deeper, but they are almost entirely absent below 3,000 metres (10,000 ft). The deepest confirmed report of a shark is a Portuguese dogfish at 3,700 metres (12,100 ft).

Relationship with Humans

Attacks

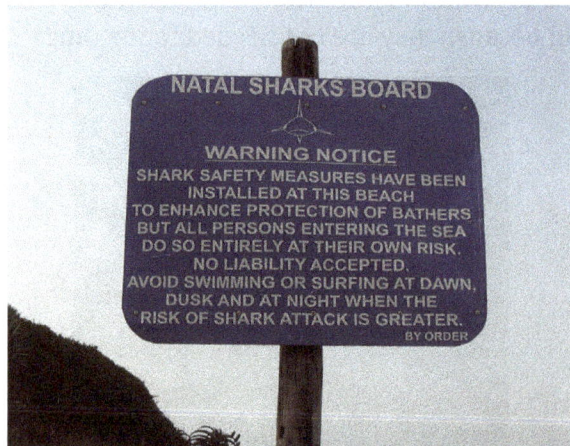

A sign warning about the presence of sharks in Salt Rock, South Africa

Snorkeler swims near blacktip reef shark. In rare circumstances involving poor visibility, blacktips may bite a human, mistaking it for prey. Under normal conditions they are harmless and shy.

In 2006 the International Shark Attack File (ISAF) undertook an investigation into 96 alleged shark attacks, confirming 62 of them as unprovoked attacks and 16 as provoked attacks. The average number of fatalities worldwide per year between 2001 and 2006 from unprovoked shark attacks is 4.3.

Contrary to popular belief, only a few sharks are dangerous to humans. Out of more than 470 species, only four have been involved in a significant number of fatal, unprovoked attacks on humans:

the great white, oceanic whitetip, tiger, and bull sharks. These sharks are large, powerful predators, and may sometimes attack and kill people. Despite being responsible for attacks on humans they have all been filmed without using a protective cage.

The perception of sharks as dangerous animals has been popularized by publicity given to a few isolated unprovoked attacks, such as the Jersey Shore shark attacks of 1916, and through popular fictional works about shark attacks, such as the *Jaws* film series. *Jaws* author Peter Benchley, as well as *Jaws* director Steven Spielberg later attempted to dispel the image of sharks as man-eating monsters.

To help avoid an unprovoked attack, humans should not wear jewelry or metal that is shiny and refrain from splashing around too much.

In Captivity

Until recently, only a few benthic species of shark, such as hornsharks, leopard sharks and catsharks, had survived in aquarium conditions for a year or more. This gave rise to the belief that sharks, as well as being difficult to capture and transport, were difficult to care for. More knowledge has led to more species (including the large pelagic sharks) living far longer in captivity, along with safer transportation techniques that have enabled long distance transportation. For a long time, the great white shark had never been successfully held in captivity for long, but in September 2004, the Monterey Bay Aquarium successfully kept a young female for 198 days before releasing her.

A whale shark in Georgia Aquarium

Most species are not suitable for home aquaria, and not every species sold by pet stores are appropriate. Some species can flourish in home saltwater aquaria. Uninformed or unscrupulous dealers sometimes sell juvenile sharks like the nurse shark, which upon reaching adulthood is far too large for typical home aquaria. Public aquaria generally do not accept donated specimens that have outgrown their housing. Some owners have been tempted to release them. Species appropriate to home aquaria represent considerable spatial and financial investments as they generally approach adult lengths of 3 feet (90 cm) and can live up to 25 years.

In Hawaii

Sharks figure prominently in Hawaiian mythology. Stories tell of men with shark jaws on their back who could change between shark and human form. A common theme was that a shark-man would warn beach-goers of sharks in the waters. The beach-goers would laugh and ignore the warnings and get eaten by the shark-man who warned them. Hawaiian mythology also includes many shark gods. Among a fishing people, the most popular of all aumakua, or deified ancestor guardians, are shark aumakua. Kamaku describes in detail how to offer a corpse to become a shark. The body transforms gradually until the kahuna can point the awe-struck family to the markings on the shark's body that correspond to the clothing in which the beloved's body had been wrapped. Such a shark aumakua becomes the family pet, receiving food, and driving fish into the family net and warding off danger. Like all aumakua it had evil uses such as helping kill enemies. The ruling chiefs typically forbade such sorcery. Many Native Hawaiian families claim such an aumakua, who is known by name to the whole community.

Kamohoali'i is the best known and revered of the shark gods, he was the older and favored brother of Pele, and helped and journeyed with her to Hawaii. He was able to assume all human and fish forms. A summit cliff on the crater of Kilauea is one of his most sacred spots. At one point he had a *heiau* (temple or shrine) dedicated to him on every piece of land that jutted into the ocean on the island of Molokai. Kamohoali'i was an ancestral god, not a human who became a shark and banned the eating of humans after eating one herself. In Fijian mythology, Dakuwaqa was a shark god who was the eater of lost souls.

In Popular Culture

In contrast to the complex portrayals by Hawaiians and other Pacific Islanders, the European and Western view of sharks has historically been mostly of fear and malevolence. Sharks are used in popular culture commonly as eating machines, notably in the *Jaws* novel and the film of the same name, along with its sequels. Sharks are threats in other films such as *Deep Blue Sea*, *The Reef*, and others, although they are sometimes used for comedic effect such as in *Finding Nemo* and the *Austin Powers* series. These comedic effects can sometimes be unintentional, as seen in *Batman: The Movie* and various Syfy channel films like *Dinoshark* and *Sharktopus*.

Sharks tend to be seen quite often in cartoons whenever a scene involves the ocean. Such examples include the *Tom and Jerry* cartoons, *Jabberjaw*, and other shows produced by Hanna-Barbera. They also are used commonly as a clichéd means of killing off a character that is held up by a rope or some similar object as the sharks swim right below them, or the character may be standing on a plank above shark infested waters.

Popular Misconceptions

A popular myth is that sharks are immune to disease and cancer, but this is not scientifically supported. Sharks have been known to get cancer. Both diseases and parasites affect sharks. The evidence that sharks are at least resistant to cancer and disease is mostly anecdotal and there have been few, if any, scientific or statistical studies that show sharks to have heightened immunity to disease. Other apparently false claims are that fins prevent cancer and treat osteoarthritis. No scientific proof supports these claims; at least one study has shown shark cartilage of no value in cancer treatment.

Conservation

The value of shark fins for shark fin soup has led to an increase in shark catches.
Usually only the fins are taken, while the rest of the shark is discarded, usually into the sea.

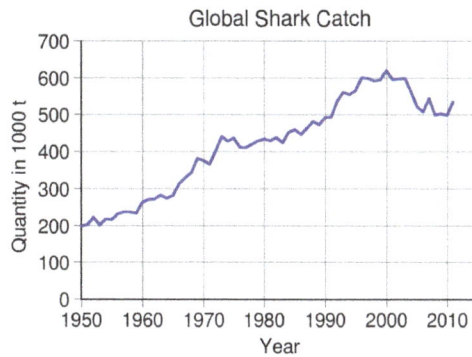

The annual shark catch has increased rapidly over the last 60 years.

A 14-foot (4.3 m), 1,200-pound (540 kg) tiger shark caught in Kāne'ohe Bay, Oahu in 1966

Fishery

It is estimated that 100 million sharks are killed by people every year, due to commercial and recreational fishing. Shark finning yields are estimated at 1.44 million metric tons for 2000, and 1.41 million tons for 2010. Based on an analysis of average shark weights, this translates into a total annual mortality estimate of about 100 million sharks in 2000, and about 97 million sharks in 2010, with a total range of possible values between 63 and 273 million sharks per year. Sharks are a common seafood in many places, including Japan and Australia. In the Australian state of Victoria, shark is the most commonly used fish in fish and chips, in which fillets are battered and deep-fried or crumbed and grilled. In fish and chip shops, shark is called flake. In India, small sharks or baby sharks (called sora in Tamil language, Telugu language) are sold in local markets. Since the flesh is not developed, cooking the flesh breaks it into powder, which is then fried in oil and spices (called sora puttu/sora poratu). The soft bones can be easily chewed. They are considered a delicacy in coastal Tamil Nadu. Icelanders ferment Greenland sharks to produce hákarl, which is widely regarded as a national dish. During a four-year period from 1996 to 2000, an estimated 26 to 73 million sharks were killed and traded annually in commercial markets.

Sharks are often killed for shark fin soup. Fishermen capture live sharks, fin them, and dump the finless animal back into the water. Shark finning involves removing the fin with a hot metal blade. The resulting immobile shark soon dies from suffocation or predators. Shark fin has become a major trade within black markets all over the world. Fins sell for about $300/lb in 2009. Poachers illegally fin millions each year. Few governments enforce laws that protect them. In 2010 Hawaii became the first U.S. state to prohibit the possession, sale, trade or distribution of shark fins. From 1996 to 2000, an estimated 38 million sharks had been killed per year for harvesting shark fins.

Shark fin soup is a status symbol in Asian countries, and is considered healthy and full of nutrients. Sharks are also killed for meat. European diners consume dogfishes, smoothhounds, catsharks, makos, porbeagle and also skates and rays. However, the U.S. FDA lists sharks as one of four fish (with swordfish, king mackerel, and tilefish) whose high mercury content is hazardous to children and pregnant women.

Sharks generally reach sexual maturity only after many years and produce few offspring in comparison to other harvested fish. Harvesting sharks before they reproduce severely impacts future populations.

The majority of shark fisheries have little monitoring or management. The rise in demand for shark products increases pressure on fisheries. Major declines in shark stocks have been recorded—some species have been depleted by over 90% over the past 20–30 years with population declines of 70% not unusual. A study by the International Union for Conservation of Nature suggests that one quarter of all known species of sharks and rays are threatened by extinction and 25 species were classified as critically endangered.

Shark Culling

A shark cull in Western Australia killed dozens of sharks in 2014, mostly tiger sharks, until it was cancelled after public protests and a decision by the Western Australia EPA; there is an ongoing "imminent threat" policy in Western Australia in which sharks that "threat" humans in the ocean can be shot and killed. This "imminent threat" policy has been criticized by senator Rachel Siewart for killing endangered sharks. From 1962 to the present, the government of Queensland has targeted and killed sharks in large numbers by using drum lines, under a "shark control" program—this program has also inadvertently killed large numbers of other animals such as dolphins. The government of New South Wales also has a program that deliberately kills sharks using nets. Kwa-zulu-Natal, an area of South Africa, has a shark-killing program using nets and drum lines—these nets and drum lines have killed turtles and dolphins, and have been criticized for killing wildlife. However it should be noted that when shark control programs are implemented on a consistent and long term basis, all of the programs have been very successful in reducing the incidence of shark attack at the protected beaches.

Other Threats

Other threats include habitat alteration, damage and loss from coastal development, pollution and the impact of fisheries on the seabed and prey species. The 2007 documentary, *Sharkwater* exposed how sharks are being hunted to extinction.

Protection

In 1991, South Africa was the first country in the world to declare Great White sharks a legally protected species.

Intending to ban the practice of shark finning while at sea, the United States Congress passed the Shark Finning Prohibition Act in 2000. Two years later the Act saw its first legal challenge in *United States v. Approximately 64,695 Pounds of Shark Fins*. In 2008 a Federal Appeals Court ruled that a loophole in the law allowed non-fishing vessels to *purchase* shark fins from fishing vessels while on the high seas. Seeking to close the loophole, the Shark Conservation Act was passed by Congress in December 2010, and it was signed into law in January 2011.

In 2003, the European Union introduced a general shark finning ban for all vessels of all nationalities in Union waters and for all vessels flying a flag of one of its member states. This prohibition was amended in June 2013 to close remaining loopholes.

In 2009, the International Union for Conservation of Nature's *IUCN Red List of Endangered Species* named 64 species, one-third of all oceanic shark species, as being at risk of extinction due to fishing and shark finning.

In 2010, the Convention on International Trade in Endangered Species (CITES) rejected proposals from the United States and Palau that would have required countries to strictly regulate trade in several species of scalloped hammerhead, oceanic whitetip and spiny dogfish sharks. The majority, but not the required two-thirds of voting delegates, approved the proposal. China, by far the world's largest shark market, and Japan, which battles all attempts to extend the convention to marine species, led the opposition. In March 2013, three endangered commercially valuable sharks, the hammerheads, the oceanic whitetip and porbeagle were added to Appendix 2 of CITES, bringing shark fishing and commerce of these species under licensing and regulation.

In 2010, Greenpeace International added the school shark, shortfin mako shark, mackerel shark, tiger shark and spiny dogfish to its seafood red list, a list of common supermarket fish that are often sourced from unsustainable fisheries. Advocacy group Shark Trust campaigns to limit shark fishing. Advocacy group Seafood Watch directs American consumers to not eat sharks.

Under the auspices of the Convention on the Conservation of Migratory Species of Wild Animals (CMS), also known as the Bonn Convention, the Memorandum of Understanding on the Conservation of Migratory Sharks was concluded and came into effect in March 2010. It was the first global instrument concluded under CMS and aims at facilitating international coordination for the protection, conservation and management of migratory sharks, through multilateral, intergovernmental discussion and scientific research.

In July 2013, New York state, a major market and entry point for shark fins, banned the shark fin trade joining seven other states of the United States and the three Pacific U.S territories in providing legal protection to sharks.

Forage Fish

Forage fish, also called prey fish or bait fish, are small pelagic fish which are preyed on by larger predators for food. Predators include other larger fish, seabirds and marine mammals. Typical ocean forage fish feed near the base of the food chain on plankton, often by filter feeding. They include particularly fishes of the family Clupeidae (herrings, sardines, shad, hilsa, menhaden, anchovies and sprats), but also other small fish, including halfbeaks, silversides, smelt such as capelin, and the goldband fusiliers pictured on the right.

Forage fish compensate for their small size by forming schools. Some swim in synchronised grids with their mouths open so they can efficiently filter plankton. These schools can become immense shoals which move along coastlines and migrate across open oceans. The shoals are concentrated fuel resources for the great marine predators. The predators are keenly focused on the shoals, acutely aware of their numbers and whereabouts, and make migrations themselves that can span thousands of miles to connect, or stay connected, with them.

The ocean primary producers, mainly contained in plankton, produce food energy from the sun and are the raw fuel for the ocean food webs. Forage fish transfer this energy by eating the plankton and becoming food themselves for the top predators. In this way, forage fish occupy the central positions in ocean and lake food webs.

The fishing industry catches forage fish primarily for feeding to farmed animals. Some fisheries scientists are expressing concern that this will affect the populations of predator fish that depend on them.

In the Oceans

Typical ocean forage fish are small, silvery schooling oily fish such as herring, anchovies and menhaden, and other small, schooling baitfish like capelin, smelts, sand lance, halfbeaks, pollock, butterfish and juvenile rockfish. Herrings are a preeminent forage fish, often marketed as sardines or pilchards.

The term "forage fish" is a term used in fisheries, and is applied also to forage species that are not true fish, but play a significant role as prey for predators. Thus invertebrates such as squid and shrimp are also referred to as "forage fish". Even the tiny shrimp-like creatures called krill, small enough to be eaten by other forage fish, yet large enough to eat the same zooplankton as forage fish, are often classified as "forage fish".

Ocean forage fish		
Anchovies	**Caribbean reef squid**	**Menhaden**
Sardines	**Shrimp**	**Northern krill**

Forage fish utilise the biomass of copepods, mysids and krill in the pelagic zone to become the dominant converters of the enormous ocean production of zooplankton. They are, in turn, central prey items for higher trophic levels. Forage fish may have achieved their dominance because of the way they live in huge, and often extremely fast cruising schools.

Though forage fish are abundant, there are relatively few species. There are more species of primary producers and apex predators in the ocean than there are forage fish.

Ocean Food Webs

Forage fish occupy central positions in the ocean food webs. The position that a fish occupies in a food web is called its trophic level (Greek *trophē* = food). The organisms it eats are at a lower trophic level, and the organisms that eat it are at a higher trophic level. Forage fish occupy middle levels in the food web, serving as a dominant prey to higher level fish, seabirds and mammals.

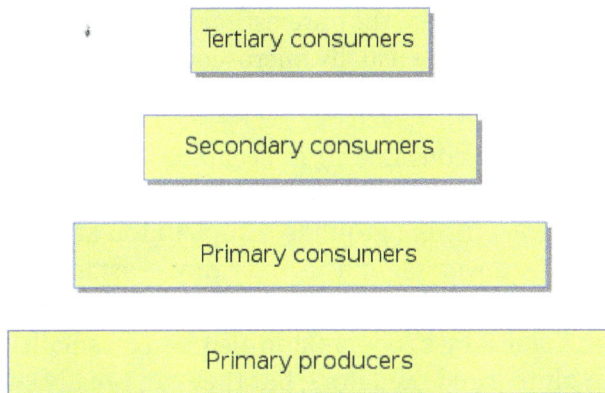

Ecological pyramids are graphical representations, along the lines of the diagram at the right, which show how biomass or productivity changes at each trophic level in an ecosystem. The first or bottom level is occupied by primary producers or autotrophs (Greek *autos* = self and *trophe* = food). These are the names given to organisms that do not feed on other organisms, but produce biomass from inorganic compounds, mostly by a process of photosynthesis.

In oceans, most primary production is performed by algae. This is a contrast to land, where most primary production is performed by vascular plants. Algae ranges from single floating cells to attached seaweeds, while vascular plants are represented in the ocean by groups such as the seagrasses. Larger producers, such as seagrasses and seaweeds, are mostly confined to the littoral zone and shallow waters, where they attach to the underlying substrate and still be within the photic zone. Most primary production in the ocean is performed by microscopic organisms, the phytoplankton.

Thus, in ocean environments, the first bottom trophic level is occupied principally by phytoplankton, microscopic drifting organisms, mostly one-celled algae, that float in the sea. Most phytoplankton are too small to be seen individually with the unaided eye. They can appear as a green discoloration of the water when they are present in high enough numbers. Since they increase their biomass mostly through photosynthesis they live in the sun-lit surface layer (euphotic zone) of the sea.

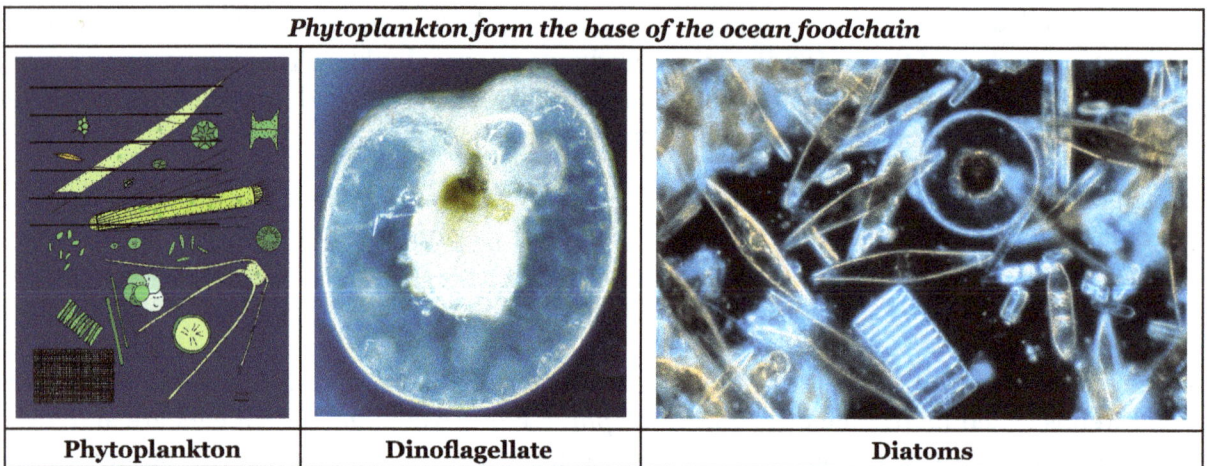

Phytoplankton form the base of the ocean foodchain		
Phytoplankton	**Dinoflagellate**	**Diatoms**

The most important groups of phytoplankton include the diatoms and dinoflagellates. Diatoms are especially important in oceans, where they are estimated to contribute up to 45% of the total ocean's primary production. Diatoms are usually microscopic, although some species can reach up to 2 millimetres in length.

The second trophic level (primary consumers) is occupied by zooplankton which feed off the phytoplankton. Together with the phytoplankton, they form the base of the food pyramid that supports most of the world's great fishing grounds. Zooplankton are tiny animals found with the phytoplankton in oceanic surface waters, and include tiny crustaceans, and fish larvae and fry (recently hatched fish). Most zooplankton are filter feeders, and they use appendages to strain the phytoplankton in the water. Some larger zooplankton also feed on smaller zooplankton. Some zooplankton can jump about a bit to avoid predators, but they can't really swim. Like phytoplankton, they float with the currents, tides and winds instead. Zooplanktons can reproduce rapidly, their populations can increase up to thirty percent a day under favourable conditions. Many live short and productive lives and reach maturity quickly.

Zooplankton form the second level in the ocean food chain

| Segmented worm | Tiny shrimp-like crustaceans |

Particularly important groups of zooplankton are the copepods and krill. These are not shown in the images above, but are discussed in more detail later. Copepods are a group of small crustaceans found in ocean and freshwater habitats. They are the biggest source of protein in the sea, and are important prey for forage fish. Krill constitute the next biggest source of protein. Krill are particularly large predator zooplankton which feed on smaller zooplankton. This means they really belong to the third trophic level, secondary consumers, along with the forage fish.

Together, phytoplankton and zooplankton make up most of the plankton in the sea. Plankton is the term applied to any small drifting organisms that float in the sea (Greek *planktos* = wanderer or drifter). By definition, organisms classified as plankton are unable to swim against ocean currents; they cannot resist the ambient current and control their position. In ocean environments, the first two trophic levels are occupied mainly by plankton. Plankton are divided into producers and consumers. The producers are the phytoplankton (Greek *phyton* = plant) and the consumers, who eat the phytoplankton, are the zooplankton (Greek *zoon* = animal).

Diet

Forage fish feed on plankton. When they are eaten by larger predators, they transfer this energy from the bottom of the food chain to the top and in this way are the central link between trophic levels.

Forage fish are usually filter feeders, meaning that they feed by straining suspended matter and food particles from water. They usually travel in large, slow moving, tightly packed schools with their mouths open. They are typically omnivorous. Their diet is usually based primarily on zooplankton, although, since they are omnivorous, they also take in some phytoplankton.

Young forage fish, such as herring, mostly feed on phytoplankton and as they mature they start to consume larger organisms. Older herrings feed on zooplankton, tiny animals that are found in oceanic surface waters, and fish larvae and fry (recently hatched fish). Copepods and other tiny crustaceans are common zooplankton eaten by forage fish. During daylight, many forage fish stay in the safety of deep water, feeding at the surface only at night when there is less chance of predation. They swim with their mouths open, filtering plankton from the water as it passes through their gills.

Ocean halfbeaks are omnivores which feed on algae, plankton, marine plants like seagrass, invertebrates like pteropods and crustaceans and smaller fishes. Some tropical species feed on animals during the day and plants at night, while others alternate summer carnivory with winter herbivory. They are in turn eaten by billfish, mackerel, and sharks.

Predators

Forage fish are the food that sustains larger predators above them in the ocean food chain. The superabundance they present in their schools make them ideal food sources for top predator fish such as tuna, striped bass, cod, salmon, barracuda and swordfish, as well as sharks, whales, dolphins, porpoises, seals, sea lions, and seabirds.

Ocean predators of forage fish

Tuna   Shark   Striped bass

Leopard seal   Dolphin

Schooling

Underwater video loop of a school of herrings migrating at high speed to their spawning grounds in the Baltic Sea.

Forage fish compensate for their small size by forming schools. These sometimes immense gatherings fuel the ocean food web. Most forage fish are pelagic fish, which means they form their schools in open water, and not on the bottom (benthic fish) or near the bottom (benthopelagic fish). They are short-lived, and go mostly unnoticed by humans, apart from an occasional support role in a documentary about a great ocean predator. While we may not pay them much attention, the great marine predators are keenly focused on them, acutely aware of their numbers and whereabouts, and make migrations that can span thousands of miles to connect with them. After all, forage fish are their food.

Herring are among the most spectacular schooling fish. They aggregate together in huge numbers. Schools have been measured at over four cubic kilometres in size, containing about four billion fish. These schools move along coastlines and traverse the open oceans. Herring schools in general have very precise arrangements which allow the school to maintain relatively constant cruising speeds. Herrings have excellent hearing, and their schools react very fast to a predator. The herrings keep a certain distance from a moving scuba diver or cruising predator like a killer whale, forming a vacuole which can look like a doughnut from a spotter plane. The intricacies of schooling is far from fully understood, especially the swimming and feeding energetics. Many hypotheses to explain the function of schooling have been suggested, such as better orientation, synchronized hunting, predator confusion and reduced risk of being found. Schooling also has disadvantages, such as excretion buildup in the breathing media and oxygen and food depletion. The way the fish array in the school probably gives energy saving advantages, though this is controversial.

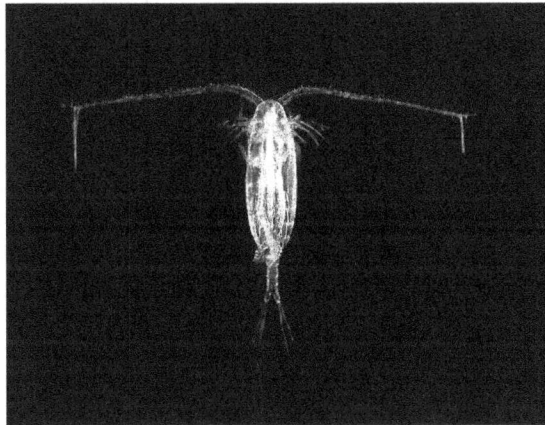

Copepod

On calm days, schools of herring can be detected at the surface a mile away by little waves they form, or from several meters at night when they trigger bioluminescence in surrounding plankton. Underwater recordings show herring constantly cruising at high speeds up to 108 cm per second, with much higher escape speeds.

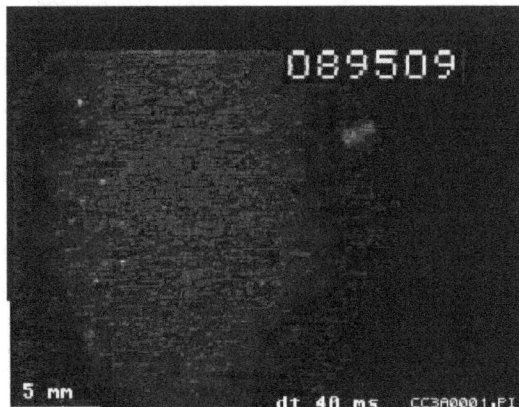

Slow motion video loop of a juvenile herring feeding on copepods.

They are fragile fish, and because of their adaptation to schooling behaviour they are rarely displayed in aquaria. Even with the best facilities aquaria can offer they become sluggish compared to their quivering energy in wild schools.

Hunting Copepods

Copepods are a group of small crustaceans found in ocean and freshwater habitats. Many species are planktonic (drifting in the ocean water), while others are benthic (living on the sea floor). Copepods are typically one millimetre (0.04 in) to two millimetres (0.08 in) long, with a teardrop shaped body. Like other crustaceans they have an armoured exoskeleton, but they are so small that this armour, and the entire body, is usually transparent.

Copepods are usually the dominant zooplankton. Some scientists say they form the largest animal biomass on the planet. The other contender is the Antarctic krill. But copepods are smaller than krill, with faster growth rates, and they are more evenly distributed throughout the oceans. This means copepods almost certainly contribute more secondary production to the world's oceans than krill, and perhaps more than all other groups of marine organisms together. They are a major item on the forage fish menu.

Copepods are very alert and evasive. They have large antennae. When they spread their antennae they can sense the pressure wave from an approaching fish and jump with great speed over a few centimeters.

Herring ram feeding on a school of copepods

Herrings are pelagic feeders. Their prey consists of a wide spectrum of phytoplankton and zooplankton, amongst which copepods are the dominant prey. Young herring usually capture small copepods by hunting them individually— they approach them from below. The (half speed) video loop at the left shows a juvenile herring feeding on copepods. In the middle of the image a copepod escapes successfully to the left. The opercula (hard bony flaps covering the gills) are spread wide open to compensate the pressure wave which would alert the copepod to trigger a jump.

If prey concentrations reach very high levels, the herrings adopt a method called "ram feeding". They swim with their mouth wide open and their opercula fully expanded. Every several feet, they close and clean their gill rakers for a few milliseconds (filter feeding). Juvenile herring hunt the copepods in synchronization: The copepods sense with their antennae the pressure-wave of an approaching herring and react with a fast escape jump. The length of the jump is fairly constant. The fish align themselves in a grid with this characteristic jump length. A copepod can dart about 80 times before it tires out. After a jump, it takes it 60 milliseconds to spread its antennae again, and this time delay becomes its undoing, as the almost endless stream of herrings allows a herring to eventually snap the copepod. A single juvenile herring could never catch a large copepod.

Migrations

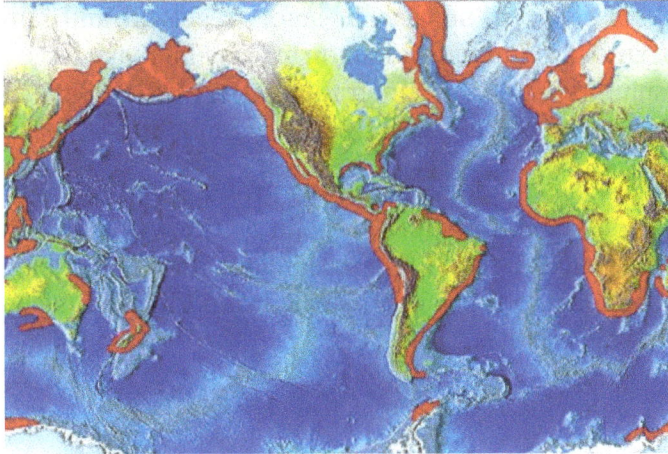

Coastal upwellings can provide plankton rich feeding grounds for forage fish.

Forage fish often make great migrations between their spawning, feeding and nursery grounds. Schools of a particular stock usually travel in a triangle between these grounds. For example, one stock of herrings have their spawning ground in southern Norway, their feeding ground in Iceland, and their nursery ground in northern Norway. Wide triangular journeys such as these may be important because forage fish, when feeding, cannot distinguish their own offspring.

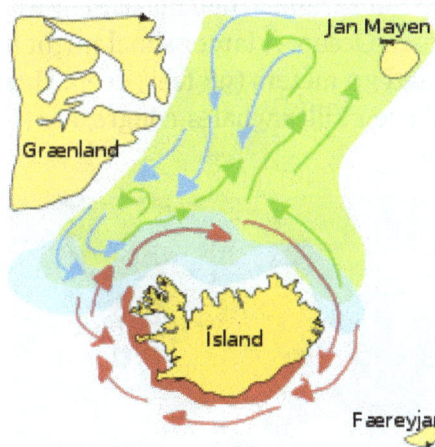

Migration of Icelandic capelin

Fertile feeding grounds for forage fish are provided by ocean upwellings. Oceanic gyres are large-scale ocean currents caused by the Coriolis effect. Wind-driven surface currents interact with these gyres and the underwater topography, such as seamounts and the edge of continental shelves, to produce downwellings and upwellings. These can transport nutrients which plankton thrive on. The result can be rich feeding grounds attractive to the plankton feeding forage fish. In turn, the forage fish themselves become a feeding ground for larger predator fish. Most upwellings are coastal, and many of them support some of the most productive fisheries in the world. Regions of notable upwelling include coastal Peru, Chile, Arabian Sea, western South Africa, eastern New Zealand and the California coast.

Capelin are a forage fish of the smelt family found in the Atlantic and Arctic oceans. In summer, they graze on dense swarms of plankton at the edge of the ice shelf. Larger capelin also eat krill and other crustaceans. The capelin move inshore in large schools to spawn and migrate in spring and summer to feed in plankton rich areas between Iceland, Greenland, and Jan Mayen. The migration is affected by ocean currents. Around Iceland maturing capelin make large northward feeding migrations in spring and summer. The return migration takes place in September to November. The spawning migration starts north of Iceland in December or January.

The diagram on the right shows the main spawning grounds and larval drift routes. Capelin on the way to feeding grounds is coloured green, capelin on the way back is blue, and the breeding grounds are red. In a paper published in 2009, researchers from Iceland recount their application of an interacting particle model to the capelin stock around Iceland, successfully predicting the spawning migration route for 2008.

Predator Attacks

Schooling forage fish are subject to constant attacks by predators. An example is the attacks that take place during the African sardine run. The African sardine run is a spectacular migration by millions of silvery sardines along the southern coastline of Africa. In terms of biomass, the sardine run could rival East Africa's great wildebeest migration.

Sardines have a short life-cycle, living only two or three years. Adult sardines, about two years old, mass on the Agulhas Bank where they spawn during spring and summer, releasing tens of thousands of eggs into the water. The adult sardines then make their way in hundreds of shoals towards the sub-tropical waters of the Indian Ocean. A larger shoal might be 7 kilometers (4.3 miles) long, 1.5 kilometers (0.93 miles) wide and 30 meters (98 feet) deep. Huge numbers of sharks, dolphins, tuna, sailfish, Cape fur seals and even killer whales congregate and follow the shoals, creating a feeding frenzy along the coastline.

Gannet

When threatened, sardines instinctively group together and create massive bait balls. Bait balls can be up to 20 meters (66 feet) in diameter. They are short lived, seldom lasting longer than 20 minutes. As many as 18,000 dolphins, behaving like sheepdogs, round the sardines into these bait balls, or herd them to shallow water (corralling) where they are easier to catch. Once rounded up, the dolphins and other predators take turns plowing through the bait balls, gorging on the fish as they sweep through. Seabirds also attack them from above, flocks of gannets, cormorants, terns and gulls. Some of these seabirds plummet from heights of 30 metres (98 feet), plunging through the water leaving vapour-like trails behind like fighter planes.

The eggs, left behind at the Agulhas Banks, drift northwest with the current into waters off the west coast, where the larvae develop into juvenile fish. When they are old enough, they aggregate into dense shoals and migrate southwards, returning to the Agulhas banks in order to restart the cycle.

Forage Fisheries

History

Medieval herring fishing in Scania (published 1555).

Herring has been known as a staple food source since 3000 B.C. In Roman times, anchovies were the base for the fermented fish sauce called *garum*. This staple of cuisine was produced in industrial quantities and transported over long distances.

Fishing for sardela or sardina (*Sardina pilchardus*) is an ongoing activity on the Croatian Adriatic coasts of Dalmatia and Istria. It traces its roots back thousands of years. The region was then largely a Venetian dominion, part of the Roman Empire. The area has always been sustained through fishing mainly sardines. Along the coast towns still promote the traditional practice of fishing by lateen sail boats for tourism and festivals.

Pilchard fishing and processing thrived in Cornwall between 1750 and 1880, after which stocks went into an almost terminal decline. Recently (2007) stocks have been improving. The industry has featured in many works of art, including Stanhope Forbes and other Newlyn School artists.

Contemporary

Traditional commercial fisheries were directed towards high value ocean predators such as cod, rockfish and tuna, rather than forage fish. As technologies developed, fisheries became so effective at locating and catching predator fish that many of the stocks collapsed. The industry compensated by turning to species lower in the food chain.

Purse seine boats encircling a school of menhaden

Commercial herring catch

In former times, forage fish were more difficult to fish profitably, and were a small part of the global marine fisheries. But modern industrial fishing technologies have enabled the removal of increasing quantities. Industrial-scale forage fish fisheries need large scale landings of fish to return profits. They are dominated by a small number of corporate fishing and processing companies.

Forage fish populations are very vulnerable when faced with modern fishing equipment. They swim near the surface in compacted schools, so they are relatively easy to locate at the surface with sophisticated electronic fishfinders and from above with spotter planes. Once located, they are scooped out of the water using highly efficient nets, such as purse seines, which remove most of the school.

Spawning patterns in forage fish are highly predictable. Some fisheries use knowledge of these patterns to harvest the forage species as they come together to spawn, removing the fish before they have actually spawned. Fishing during spawning periods or at other times when forage fish amass in large numbers can also be a blow to predators. Many predators, such as whales, tuna and sharks, have evolved to migrate long distances to specific sites for feeding and breeding. Their survival hinges on their finding these forage schools at their feeding grounds. The great ocean predators find that, no matter how they are adapted for speed, size, endurance or stealth, they are on the losing side when faced with the machinery of contemporary industrial fishing.

Altogether, forage fish account for 37 percent (31.5 million tonnes) of all fish taken from the world's oceans each year. However, because there are fewer species of forage fish compared to predator fish, forage species fisheries are the largest in the world. Seven of the top ten fisheries target forage fish. The total world catch of herrings, sardines and anchovies alone in 2005 was 22.4 million tonnes, 24 percent of the total world catch.

The Peruvian anchoveta fishery is now the biggest in the world (10.7 million tonnes in 2004), while the Alaskan pollock fishery in the Bering Sea is the largest single species fishery in the world (3 million tonnes). The Alaskan pollock is said to be the largest remaining single species source of palatable fish in the world. However, the biomass of pollock has declined in recent years, perhaps spelling trouble for both the Bering Sea ecosystem and the commercial fishery it supports. Acoustic surveys by NOAA indicate that the 2008 pollock population is almost 50 percent lower than last year's survey levels. Some scientists think this decline in Alaska pollock could repeat the collapse experienced by Atlantic cod, which could have negative consequences for the entire Bering Sea ecosystem. Salmon, halibut, endangered Steller sea lions, fur seals, and humpback whales eat pollock and depend on healthy populations to sustain themselves.

Use as Animal Feed

Eighty percent of the forage fish caught are fed to animals, in large part due to the high content of beneficial long chain omega-3 fatty acids in their flesh. Ninety percent is processed into fishmeal and fish oil. Of this, 46 percent was fed to farmed fish, 24 percent to pigs, and 22 percent to poultry (2002). Six times the weight of forage fish is fed to pigs and poultry alone than the entire seafood consumption of the U.S. market. One of the most promising alternatives to fish oil as a source of long chain omega-3 fatty acids and certain amino acids is algal oil from microalgae, the original source of these fatty acids in forage fish.

According to Turchini and De Silva (2008), another 2.5 million tonnes of the annual forage fish catch is consumed by the global cat food industry. In Australia, pet cats eat 13.7 kilograms of fish a year compared to the 11 kilograms eaten by the average Australian. The pet food industry is increasingly marketing premium and super-premium products, when different raw materials, such as the by-products of the fish filleting industry, could be used instead.

Environmental Issues

In 2008 the Sea Around Us Project completed a nine-year study of forage fish led by the fisheries scientists Jacqueline Alder and Daniel Pauly. They concluded that:

1. The composition of landings of forage fish fisheries have changed over the past 50 years with the trophic level of fish used in fishmeal increasing over the past 20 years.

2. Our understanding of the role of forage fish in marine ecosystem and the impact of fishing is still limited.

3. Landing of forage fish peaked by the 1970s, and these high levels are highly unlikely in the future, even if fisheries are managed sustainably.

4. The consumption of forage fish by seabirds and marine mammals is not likely to be onerous to fisheries, except in a few localized areas. By contrast, fisheries, by reducing the bio-

mass of small pelagics, might pose a threat to these predators, particularly to those species for which stocks have been heavily depleted by human exploitation in the past.

5. Some forage fish species are consumed by many people with consumption patterns changing over the last 20 years.

6. Aquaculture continues to increase its consumption of fishmeal and fish oil.

In 2015 sardine populations crashed along the west coast of the United States, causing the fishery to close early and remain closed through the 2015–2016 season. A key reason for the population crash, alongside climate change, was overfishing due to the demand of fish meals and fish oil used in feed for aquaculture and for human nutritional supplements. In an effort to provide some relief from the pressure put on forage fish populations, the World Bank along with the University of Arizona, Monterey Bay Aquarium and the New England Aquarium has sponsored a competition called the F3 (Fish-Free Feed) Challenge, which will award $200,000 to the most successful fish feed manufacturer who develops aquaculture feeds not made from fish.

In Lakes and Rivers

Forage fish also inhabit freshwater habitats, such as lakes and rivers, where they serve as food for larger freshwater predators. Usually smaller than 15 centimetres (6 in) in length, these small bait fish make up most of the fish found in lakes and rivers. The minnow family alone, consisting of minnows, chubs, shiners and daces, consists of more than fifty species. Other freshwater forage fish include suckers, killifish, shad, bony fish as well as fish of the sunfish family, excluding black basses and crappie, and smaller species of the carp family. There are also anadromous forage fish, such as eulachon.

Freshwater forage fish

Golden shiner

Killifish

Southern redbelly dace

Chinese minnow

Twaite shad

Swarm of carp

Within any fresh or saltwater ecosystem, there will always be both desirable and undesirable fishes, and this varies from country to country, and often from region to region within a country. Sport fishermen divide freshwater predators of forage fish into those:

- which have a good fighting ability and are good to eat, called sport (or game) fish.

- the other less desirable fish, called rough fish in North America and coarse fish in Britain

Rough or coarse fish usually refers to fish that are not commonly eaten, not sought after for sporting reasons, or have become invasive species reducing the populations of desirable fish. They compete for forage fish with the more popular sport fish. They are often regarded as a nuisance, and are not usually protected by game laws. Forage fish generally are not considered rough or coarse fish because of their usefulness as bait.

The term *rough fish* is used by U.S. state agencies and anglers to describe undesirable predator fish. In North America, anglers fish for salmon, trout, bass, pike, catfish, walleye and muskellunge. The smallest fish are called panfish, because they can fit in a standard cooking pan. Some examples are crappies, rock bass, perch, bluegill and sunfish.

Freshwater predators of forage fish

Brook trout

Black crappie

Macquarie perch

Rainbow trout

Pink salmon

Channel catfish

The term *coarse fish* originated in the United Kingdom in the early 19th century. Prior to that time, recreational fishing was the sport of the gentry, who angled for trout and salmon which they called "game fish". Fish other than game fish were disdained as "coarse fish". These days, "game fish" refers to Salmonids (other than grayling) — that is, salmon, trout and char. Coarse fish are made up mostly of the larger species of Cyprinids (carp, roach, bream) as well as pike, catfish, gar and lamprey. Coarse fish are no longer disdained; indeed, fishing for coarse fish has become a popular pastime.

Bait and Feeder Fish

Forage fish are sometimes referred to as *bait fish* or *feeder fish*. Bait fish is a term used particularly by recreational fishermen, although commercial fisherman also catch fish to bait longlines and traps. Forage fish is a fisheries term, and is used in the context of fisheries. Bait fish, by contrast, are fish that are caught by humans to use as bait for other fish. The terms overlap in the sense that most bait fish are also forage fish, and vice versa. Feeder fish is a term used particularly in the context of fish aquariums. It refers essentially to the same concept as forage fish, small fish that are eaten by larger fish, but the term is adapted to the particular requirements of working with fish in aquariums.

Timeline

- 2006: The U.S. National Coalition for Marine Conservation asks U.S. fishery managers to put "Forage First!". Their campaign was launched with the publication of their report, *Taking the Bait: Are America's Fisheries Out-competing Predators for their Prey?*, available at cost to the U. S. fishing industry, encouraging fishery managers to protect predator–prey relationships as a first step toward an ecosystem based approach to fishery management.

- 2009: The international Lenfest Forage Fish Task Force is established to develop workable management plans for tackling the depletion of forage fish.

- 2015: sardine populations crashed along the west coast of the United States.

Recent Reports

- Pikitch E and 12 others (2012) Little Fish, Big Impact: Managing a Crucial Link in Ocean Food Webs *Lenfest Ocean Program*, Washington, DC. Summary and other materials available on the Lenfest Ocean Program website.

Oily Fish

Large open-water fish, like this Atlantic bluefin tuna, are oily fish.

Most small forage fish, like these schooling anchovies, are also oily fish.

Oily fish have oil in their tissues and in the belly cavity around the gut. Their fillets contain up to 30% oil, although this figure varies both within and between species. Examples include small forage fish, such as sardines, herring and anchovies, and other larger pelagic fish, such as salmon, trout, tuna and mackerel.

Oily fish can be contrasted with white fish, which contain oil only in the liver, and much less overall than oily fish. Examples of white fish are cod, haddock and flatfish. White fish are usually demersal fish which live on or near the seafloor, whereas oily fish are pelagic, living in the water column away from the bottom.

Oily fish meat is a good source of vitamins A and D, and is rich in omega-3 fatty acids (white fish also contain these nutrients but at a much lower concentration). For this reason the consumption of oily fish rather than white fish can be more beneficial to humans, particularly concerning cardiovascular diseases; however, oily fish are known to carry higher levels of contaminants (such as mercury or dioxin) than whitefish. Among other benefits, studies suggest that the omega-3 fatty acids in oily fish may help improve inflammatory conditions such as arthritis.

Health Benefits

Oily fish fillet (salmon – bottom) contrasted with a white fish fillet (halibut – top)

Dementia

A 1997 study published in *Annals of Neurology* followed 5,386 elderly participants in Rotterdam. It found that fish consumption decreased the risk of dementia. However, the 2.1-year average follow-up was less than the three years dementia commonly affects people prior to diagnosis. Thus, the study was unclear as to whether fish consumption protected against dementia, or if dementia prevented the participants from wanting more fish.

French research published in 2002 in the *British Medical Journal* (BMJ) followed 1,674 elderly residents of southern France for seven years, studying their consumption of meat versus seafood and the presence of dementia symptoms. The conclusion was that people who ate fish at least once a week had a significantly lower risk of being diagnosed with dementia over a seven-year period.

This study reinforced the Annals of Neurology findings. Because of the longer term, the BMJ study provided stronger evidence of a genuine protective effect. There was a possible confounding factor in that individuals with higher education have both a lower risk of dementia and higher consumption of fish.

Cardiovascular Health

Consuming 200–400 g of oily fish twice per week may also help prevent sudden death due to myocardial infarction by preventing cardiac arrhythmia. The eicosapentaenoic acid found in fish oils appears to dramatically reduce inflammation through conversion within the body to resolvins, with beneficial effects for the cardiovascular system and arthritis.

Recommended Consumption

In 1994, the UK Committee on Medical Aspects of Food and Nutrition Policy recommended that people eat at least two portions of fish per week, one of which should be oily fish.

In 2004 the UK Food Standards Agency published advice on the recommended minimum and maximum quantities of oily fish to be eaten per week, to balance the beneficial qualities of the omega-3 fatty acids against the potential dangers of ingesting polychlorinated biphenyls and dioxins. It reiterated the 1994 guideline of two portions of fish per week including one portion of oily fish, but advised eating no more than four portions per week, and no more than two portions for people who are pregnant, may become pregnant or who are breastfeeding.

The United States Environmental Protection Agency's (EPA) Exposure Reference Dose (RfI) for MeHg is 0.1 micrograms per kg body weight per day. The corresponding limit of blood mercury is 5.8 micrograms per liter. The restrictions apply to certain oily fish – "marlin, swordfish, shark and, to a lesser extent, tuna" The recommendations on maximum consumption of oily fish were up to four portions (1 portion = 140g, or approx 4.9 ounces) a week for men, boys, and women past childbearing age, and up to two portions a week for women of childbearing age, including pregnant and breastfeeding women, and girls. There is no recommended limit on the consumption of white fish.

The EPA and 2007 U.S. Department of Agriculture guidelines sets a limit only on consumption of fatty fish with greater than one part per million of methylmercury, specifically tilefish, king mackerel, shark and swordfish. There are limits, however, for nursing/pregnant women and children under the age of six. This population should completely avoid fish with high risk of mercury contamination (those listed above) and limit consumption of moderate and low-mercury fish to 12 ounces or fewer per week. Albacore tuna should be limited to six ounces or less per week.

Omega-3 Content

Concerns about contamination, diet or supply have led to investigation of plant sources of omega-3 fatty acids, notably flax, hempseed and perilla oils. Lactating women who supplemented their diet with flaxseed oil showed increases in blood and breastmilk concentration of alpha-linolenic acid and eicosapentaenoic acid but no changes to concentrations of docosahexaenoic acid.

Demersal Fish

Bluespotted ribbontail ray resting on the seafloor

Rhinogobius flumineus swim on the bed of rivers

Demersal fish live and feed on or near the bottom of seas or lakes (the demersal zone). They occupy the sea floors and lake beds, which usually consist of mud, sand, gravel or rocks. In coastal waters they are found on or near the continental shelf, and in deep waters they are found on or near the continental slope or along the continental rise. They are not generally found in the deepest waters, such as abyssal depths or on the abyssal plain, but they can be found around seamounts and islands. The word *demersal* comes from the Latin *demergere*, which means *to sink*.

Demersal fish are bottom feeders. They can be contrasted with pelagic fish which live and feed away from the bottom in the open water column. Demersal fish fillets contain little fish oil (one to four percent), whereas pelagic fish can contain up to 30 percent.

Types

Demersal fish can be divided into two main types: strictly benthic fish which can rest on the sea floor, and benthopelagic fish which can float in the water column just above the sea floor.

Benthopelagic fish have neutral buoyancy, so they can float at depth without much effort, while strictly benthic fish are more dense, with negative buoyancy so they can lie on the bottom without any effort. Most demersal fish are benthopelagic.

Benthic flatfish and benthopelagic cod on a shore – Jan van Kessel senior, 1626–1679

As with other bottom feeders, a mechanism to deal with substrate is often necessary. With demersal fish the sand is usually pumped out of the mouth through the gill slit. Most demersal fish exhibit a flat ventral region so as to more easily rest their body on the substrate. The exception may be the flatfish, which are laterally depressed but lie on their sides. Also, many exhibit what is termed an "inferior" mouth, which means that the mouth is pointed downwards; this is beneficial as their food is often going to be below them in the substrate. Those bottom feeders with upward-pointing mouths, such as stargazers, tend to seize swimming prey.

Benthic Fish

Benthic fish, sometimes called groundfish, are denser than water, so they can rest on the sea floor. They either lie-and-wait as ambush predators, maybe covering themselves with sand or otherwise camouflaging themselves, or move actively over the bottom in search for food. Benthic fish which can bury themselves include dragonets, flatfish and stingrays.

Flatfish are an order of ray-finned benthic fishes which lie flat on the ocean floor. Examples are flounder, sole, turbot, plaice, and halibut. The adult fish of many species have both eyes on one side of the head. When the fish hatches, one eye is located on each side of its head. But as the fish grows from the larval stage, one eye migrates to the other side of the body as a process of metamorphosis. The flatfish then changes its habits, and camouflages itself by lying on the bottom of the ocean floor with both eyes facing upwards. The side to which one eye migrates depends on the species; with some species both eyes are ultimately on the left side, whereas with other species the eyes are on the right.

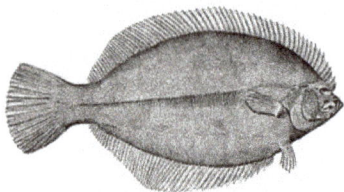

Flounder have both eyes on one side of their head.

Some flatfish can camouflage themselves on the ocean floor.

Bluespotted ribbontail rays migrate in schools onto shallow sands to feed on mollusks, shrimps, crabs and worms.

Flounder ambush their prey, feeding at soft muddy area of the sea bottom, near bridge piles, docks, artificial and coral reefs. Their diet consists mainly of fish spawn, crustaceans, polychaetes and small fish.

The great hammerhead swings its head in broad angles over the sea floor to pick up the electrical signatures of stingrays buried in the sand. It then uses its "hammer" to pin down the stingray.

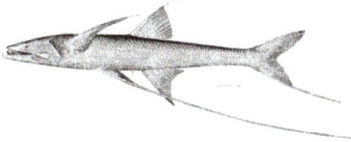

The tripodfish (*Bathypterois grallator*), a species of spiderfish, uses its fin extensions to "stand" on the bottom.

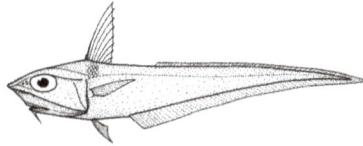

Gargoyle fish

Pacific hagfish resting on bottom. Hagfish coat themselves and any dead fish they find with noxious slime making them inedible to other species.

Some fishes don't fit into the above classification. For example, the family of nearly blind spiderfishes, common and widely distributed, feed on benthopelagic zooplankton. Yet they are strictly benthic fish, since they stay in contact with the bottom. Their fins have long rays they use to "stand" on the bottom while they face the current and grab zooplankton as it passes by.

The bodies of benthic fish are adapted for ongoing contact with the sea floor. Swimbladders are usually absent or reduced, and the bodies are usually flattened in one way or another. Following Moyle and Cech (2004) they can be divided into five overlapping body shapes:

Body types of benthic fish

Bottom rovers

Bottom rovers "have a rover-predator-like body, except that the head tends to be flattened, the back humped, and the pectoral fins enlarged. North American catfish with large mouths at the end of the snout, small armoured catfish with small mouths beneath the snout, and sturgeons, with fleshy protusible lips located well below the snout that are used to suck plant and animal matter off the bottom."

Bottom clingers

Bottom clingers "are mainly small fish with flattened heads, large pectoral fins, and structures (usually modified pelvic fins) that enable them to adhere to the bottom. Such structures are handy in swift streams, or intertidal areas with strong currents. The simplest arrangement is possessed by sculpins, which use their small, closely spaced pelvic fins, as antiskid devices. However, other families of fishes, such as gobies, and clingfishes have evolved suction cups."

Bottom hiders

Bottom hiders "are similar in many ways to bottom clingers. but they lack the clinging devices and tend to have more elongate bodies and smaller heads. These forms usually live under rocks or in crevices or lie quietly on the bottom in still waters. The darters of North American streams are in the category, as are many blennies."

Flatfish

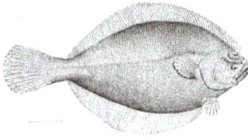

Flatfish "have the most extreme morphologies of the bottom fish. Flounders are essentially deep-bodied fish which live with one side on the bottom. In these fish, the eye on the downward side migrates during development to the upward side, and the mouth often assumes a peculiar twist to enable bottom feeding. In contrast, skates and rays are flattened dorsoventrally, and mostly move about by flapping their extremely large pectoral fins. Not only is the mouth completely ventral on these fish, the main water intakes for respiration (the spiracles) are located on the top of the head."

Rattails

Rattail-shaped fish "have bodies that begin with large pointy-snouted heads and large pectoral fins and end in long pointed rat-like tails. These fish are almost all bottom-dwelling (benthic) inhabitants of the deep sea, but exactly why this peculiar morphology is so popular among them is poorly understood. The fish live by scavenging and preying on benthic invertebrates. Examples are the grenadiers, brotulas (pictured), and chimaeras."

Benthopelagic Fish

The sluggish bathydemersal false catshark, shown here at a depth of 1,200 meters, has an enormous oil-filled liver which lets it hover off the continental slope at near-neutral buoyancy. It feeds on cephalopods, cutthroat eels, grenadiers, snake mackerel, and lanternsharks.

Benthopelagic fish inhabit the water just above the bottom, feeding on benthos and zooplankton. Most demersal fish are benthopelagic.

Deep sea benthopelagic teleosts all have swimbladders. The dominant species, rattails and cusk eels, have considerable biomass. Other species include deep sea cods (morids), deep sea eels, halosaurs and notacanths.

Benthopelagic sharks, like the deep sea squaloid sharks, achieve neutral buoyancy with the use of large oil-filled livers. Sharks adapt well to fairly high pressures. They can often be found on slopes down to about 2000 metres, scavenging on food falls such as dead whales. However, the energy demands of sharks are high, since they need to swim constantly and maintain a large amount of oil for buoyancy. These energy needs cannot be met in the extreme oligotrophic conditions that occur at great depths.

Shallow water stingrays are benthic, and can lie on the bottom because of their negative buoyancy. Deep sea stingrays are benthopelagic, and like the squaloids have very large livers which give them neutral buoyancy.

Benthopelagic fish can be divided into flabby or robust body types. Flabby benthopelagic fishes are like bathypelagic fishes; they have a reduced body mass, and low metabolic rates, expending minimal energy as they lie and wait to ambush prey. An example of a flabby fish is the cusk-eel *Acanthonus armatus*, a predator with a huge head and a body that is 90 percent water. This fish has the largest ears (otoliths) and the smallest brain in relation to its body size of all known vertebrates.

Deepwater benthopelagic fish are robust, muscular swimmers that actively cruise the bottom searching for prey. They often live around features, such as seamounts, which have strong currents. Commercial examples are the orange roughy and Patagonian toothfish.

Habitats

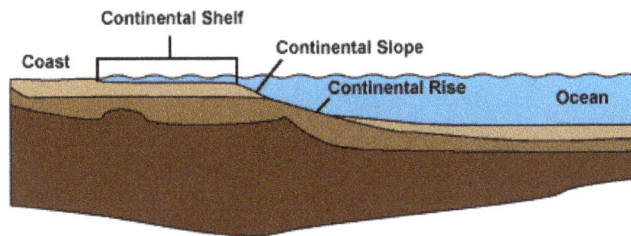

Profile illustrating the shelf, slope and rise

The edge of the continental shelf marks the boundary where the shelf gives way to, and then gradually drops into abyssal depths. This edge marks the boundary between coastal, relatively shallow, benthic habitats, and the deep benthic habitats. Coastal demersal fishes live on the bottom of inshore waters, such as bays and estuaries, and further out, on the floor of the continental shelf. Deep water demersal fish live beyond this edge, mostly down the continental slopes and along the continental rises which drop to the abyssal plains. This is the continental margin, constituting about 28% of the total oceanic area. Other deep sea demersal fish can also be found around seamounts and islands.

The term *bathydemersal fish* is sometimes used instead of "deep water demersal fish". *Bathydemersal* refers to demersal fish which live at depths greater than 200 metres.

The term *epibenthic* is also used to refer to organism that live on top of the ocean floor, as opposed to those that burrow into the seafloor substrate. However the terms *mesodemersal*, *epidemersal*, *mesobenthic* and *bathybenthic* are not used.

Coastal

Coastal demersal fish are found on or near the seabed of coastal waters between the shoreline and the edge of the continental shelf, where the shelf drops into the deep ocean. Since the continental shelf is generally less than 200 metres deep, this means that coastal waters are generally epipelagic. The term includes demersal reef fish and demersal fish that inhabit estuaries, inlets and bays.

Triggerfish use a jet of water to un-
cover sand dollars buried in sand.

The mangrove jack eats crus-
taceans

Many puffer fish species
crush the shells of molluscs

Young mangrove jacks, a sought after eating and sport fish, dwell in estuaries around mangrove roots, fallen trees, rock walls, and any other snag areas where smaller prey reside for protection. When they mature, they migrate into open waters, sometimes hundreds of kilometres from the coast to spawn.

The stargazer *Uranoscopus sulphureus*.

Stargazers are found worldwide in shallow waters. They have eyes on top of their heads and a large upward-facing mouth. They bury themselves in sand, and leap upwards to ambush benthopelagic fish and invertebrates that pass overhead. Some species have a worm-shaped lure growing out of the floor of the mouth, which they wiggle to attract prey. Stargazers are venomous and can deliver electric shocks. They have been called "the meanest things in creation."

Other examples of coastal demersal fish are cod, plaice, monkfish and sole.

Deep Water

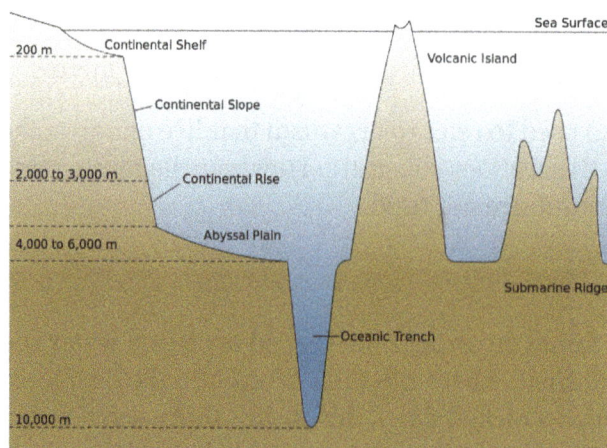
Cross-section of an ocean basin. Note significant vertical exaggeration.

Deep water demersal fish occupy the benthic regions beyond the continental margins.

On the continental slope, demersal fishes are common. They are more diverse than coastal demersal fish, since there is more habitat diversity. Further out are the abyssal plains. These flat, featureless regions occupy about 40 percent of the ocean floor. They are covered with sediment but largely devoid of benthic life (benthos). Deep sea benthic fishes are more likely to associate with canyons or rock outcroppings among the plains, where invertebrate communities are established. Undersea mountains (seamounts) can intercept deep sea currents, and cause productive upwellings which support benthic fish. Undersea mountain ranges can separate underwater regions into different ecosystems.

Rattails and brotulas are common, and other well-established families are eels, eelpouts, hagfishes, greeneyes, batfishes and lumpfishes.

The bodies of deep water demersal fishes are muscular with well developed organs. In this way they are closer to mesopelagic fishes than bathypelagic fishes. In other ways, they are more variable. Photophores are usually absent, eyes and swimbladders range from absent to well developed. They vary in size, and larger species, greater than one metre, are not uncommon.

Giant grenadier, an elongate deep water demersal fish with large eyes and well-developed lateral lines

Deep sea demersal fish are usually long and narrow. Many are eels or shaped like eels. This may be because long bodies have long lateral lines. Lateral lines detect low-frequency sounds, and some demersal fishes have muscles that drum such sounds to attract mates. Smell is also important, as indicated by the rapidity with which demersal fish find traps baited with bait fish.

The main diet of deep sea demersal fish is invertebrates of the deep sea benthos and carrion. Smell, touch and lateral line sensitivities seem to be the main sensory devices for locating these.

Like coastal demersal fish, deep sea demersal fish can be divided into benthic fish and benthopelagic fish, where the benthic fish are negatively buoyant and benthopelagic fish are neutrally buoyant.

The availability of plankton for food diminishes rapidly with depth. At 1000 metres, the biomass of plankton is typically about 1 percent of that at the surface, and at 5000 metres about 0.01 percent. Given there is no sunlight, energy enters deep water zones as organic matter. There are three main ways this happens. Firstly, organic matter can move into the zone from the continental landmass, for example, through currents that carry the matter down rivers, then plume along the continental shelf and finally spill down the continental slope. Other matter enters as particulate matter raining down from the overhead water column in the form of marine snow, or as sinking overhead plant material such as eelgrass, or as "large particles" such as dead fish and whales sinking to the bottom. A third way energy can arrive is through fish, such as vertically migrating mesopelagic fishes that can enter into the demersal zone as they ascend or descend. The demersal fish and invertebrates consume organic matter that does arrive, break it down and recycle it. A consequence of these energy delivery mechanisms is that the abundance of demersal fish and invertebrates gradually decrease as the distance from continental shorelines increases.

Although deep water demersal fish species are not generally picky about what they eat, there is still some degree of specialisation. For example, different fish have different mouth sizes, which determines the size of the prey they can handle. Some feed mostly on benthopelagic organisms. Others fed mostly on epifauna (invertebrates on top of the seafloor surface, also called *epibenthos*), or alternatively on infauna (invertebrates that burrow into the seafloor substrate). Infauna feeders can have considerable sediment in their stomachs. Scavengers, such as snubnosed eels and hagfish, also eat infauna as a secondary food source.

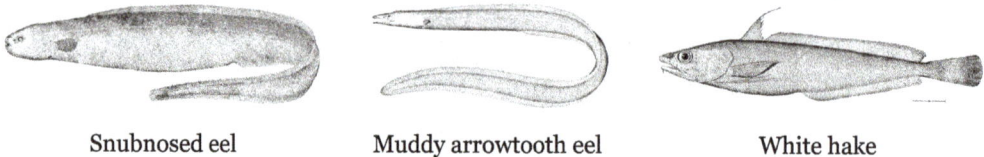

| Snubnosed eel | Muddy arrowtooth eel | White hake |

Some feed on carrion. Cameras show that when a dead fish is placed on the bottom, vertebrate and invertebrate scavengers appear very quickly. If the fish is large, some scavengers burrow in and eat it from the inside out. Some fish, such as grenadiers, also appear and start feeding on the scavenging inverebrates and amphipods. Other specialization is based on depth distribution. Some of the more abundant upper continental slope fish species, such as cutthroat eel and longfinned hake, mainly feed on epipelagic fish. But generally, the most abundant deep water demersal fish species feed on invertebrates.

At great depths, food scarcity and extreme pressure limits the ability of fish to survive. The deepest point of the ocean is about 11,000 metres. Bathypelagic fishes are not normally found below 3,000 metres. It may be that extreme pressures interfere with essential enzyme functions.

The deepest-living fish known, the strictly benthic *Abyssobrotula galatheae*, eel-like and blind, feeds on benthic invertebrates. A living example was trawled from the bottom of the Puerto Rico Trench in 1970 from a depth of 8,370 metres (27,453 ft).

In 2008, a shoal of 17 hadal snailfish, a species of deep water snailfish, was filmed by a UK-Japan team using remote operated landers at depths of 7.7 km (4.8 mi) in the Japan Trench in the Pacific. The fish were 30 centimetres long (12 in), and were darting about, using vibration sensors on their nose to catch shrimps. The team also reported that the appearance of the fish, unlike that of most deep sea fish, was surprisingly "cute", and that they were surprised by how active the fish were at these depths.

Demersal Fisheries

Most demersal fish of commercial or recreational interest are coastal, confined to the upper 200 metres. Commercially important demersal food fish species include flatfish, such as flounder, sole, turbot, plaice, and halibut. Also important are cod, hake, redfish, haddock, bass, congers, sharks, rays and chimaeras.

The following table shows the world capture production of some groups of demersal species in tonnes.

Capture production by groups of species in tonnes								
Group	1999	2000	2001	2002	2003	2004	2005	
Cods, hakes, haddocks	9,431,141	8,695,910	9,304,922	8,474,044	9,385,328	9,398,780	8,964,873	
Flounders, halibuts, soles	956,926	1,009,253	948,427	915,177	917,326	862,162	900,012	
Other demersal fishes	2,955,849	3,033,384	3,008,283	3,062,222	3,059,707	3,163,050	2,986,081	Demersal fish output in 2005

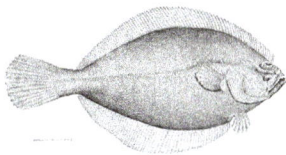

American plaice are usually found between 90 and 250 metres (but have been found at 3000 m). They feed on small fishes and invertebrates.

Atlantic cod are usually found between 150 and 200 metres, they are omnivorous and feed on invertebrates and fish, including young cod.

Grouper are ambush predators with a powerful sucking system that sucks their prey in from a distance.

Black sea bass inhabit US coasts from Maine to NE Florida and the eastern Gulf of Mexico, and are most abundant off the waters of New York. They are found in inshore waters (bays and sounds) and offshore in waters up to a depth of 130 m (425'). They spend most of their time close to the sea floor and are often congregated around bottom formations such as rocks, man-made reefs, wrecks, jetties, piers, and bridge pilings. Black sea bass are sought after recreational and commercial fish, and have been overfished.

Grouper are often found around reefs. They have stout bodies and large mouths. They are not built for long-distance or fast swimming. They can be quite large, and lengths over a meter and weights up to 100 kg are not uncommon. They swallow prey rather than biting pieces off it. They do not have many teeth on the edges of their jaws, but they have heavy crushing tooth plates inside the pharynx. They lie in wait, rather than chasing in open water. They are found in areas of hard or consolidated substrate, and use structural features such as ledges, rocks, and coral reefs (as well as artificial reefs like wrecks and sunken barges) as their habitat. Their mouth and gills form a powerful sucking system that sucks their prey in from a distance. They also use their mouth to dig into sand to form their shelters under big rocks, jetting it out through their gills. Their gill muscles are so powerful that it is nearly impossible to pull them out of their cave if they feel attacked and extend those muscles to lock themselves in. There is some research indicating that roving coral groupers (*Plectropomus pessuliferus*) sometimes cooperate with giant morays in hunting.

Deepwater benthopelagic fish are robust, muscular swimmers that actively cruise the bottom searching for prey. They often live around features, such as seamounts, which have strong currents. Commercial examples are the orange roughy and Patagonian toothfish. Because these fish were once abundant, and because their robust bodies are good to eat, these fish have been commercially harvested.

The Patagonian toothfish is a robust benthopelagic fish

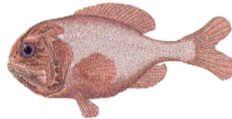

So is the orange roughy

The blue grenadier (hoki), a deep water demersal fish, is subjected to a large sustainable fishing industry in New Zealand.

Conservation Status

Major demersal fishery species in the North Sea such as cod, plaice, monkfish and sole, are listed by the ICES as "outside safe biological limits."

- The True Sole *solea solea* is sufficiently broadly distributed that it is not considered a threatened species; however, overfishing in Europe has produced severely diminished populations, with declining catches in many regions. For example, the western English Channel and Irish Sea sole fisheries face potential collapse according to data in the UK Biodiversity Action Plan.

- Sole, along with the other major bottom-feeding fish in the North Sea such as cod, monkfish, and plaice, is listed by the ICES as "outside safe biological limits." Moreover, they are growing less quickly now and are rarely older than six years, although they can reach forty. World stocks of large predatory fish and large ground fish such as sole and flounder were estimated in 2003 to be only about 10% of pre-industrial levels. According to the World Wildlife Fund in 2006, "of the nine sole stocks, seven are overfished with the status of the remaining two unknown." Data is insufficient to assess the remaining stocks; however, landings for all stocks are at or near historical lows."

- World stocks of large predatory fish and large ground fish such as sole and flounder were estimated in 2003 to be only about 10% of pre-industrial levels, largely due to overfishing. Most overfishing is due to the extensive activities of the fishing industry. Current research indicate that the flounder population could be as low as 15 million due to heavy overfishing and industrial pollution along the Gulf of Mexico surrounding the coast of Texas.

- Seafood Watch have placed on their list of seafood that sustainability-minded consumers should avoid the following demersal fish: sturgeon (imported wild), Chilean seabass, cod (Atlantic, imported Pacific), flounder (Atlantic), halibut (Atlantic), sole (Atlantic), grouper, monkfish, orange roughy, demersal shark, red snapper and tilapia (Asia farmed).

References

- Jøn, A. Asbjørn; Aich, Raj S. (2015). "Southern shark lore forty years after Jaws: The positioning of sharks within Murihiku, New Zealand". Australian Folklore: A Yearly Journal of Folklore Studies. University of New England (30)

- Stephenson, S. A. "The Distribution of Pacific Salmon ("Oncorhynchus" spp.) in the Canadian Western Arctic" (PDF). Retrieved 1 September 2013

- Finkelstein JB (2005). "Sharks do get cancer: few surprises in cartilage research". Journal of the National Cancer Institute. 97 (21): 1562–3. PMID 16264172. doi:10.1093/jnci/dji392

- Wilkinson, Charles (2000). Messages from Frank's Landing: A Story of Salmon, Treaties, and the Indian Way. University of Washington Press. ISBN 0295980117. OCLC 44391504

- McGrath, Susan. "Spawning Hope". Audubon Society. Archived from the original on 27 September 2007. Retrieved 17 November 2006

- Francois, C.A.; Connor, S. L.; Bolewicz, L. C.; Connor, W. E. (1 January 2003). "Supplementing lactating women with flaxseed oil does not increase docosahexaenoic acid in their milk". American Journal of Clinical Nutrition. 77 (1): 226–233. PMID 12499346. Retrieved 26 July 2007

- Nadel., Foley, Dana (2005-01-01). Atlas of pacific salmon : the first map-based status assessment of salmon in the North Pacific. California University Press. ISBN 0520245040. OCLC 470376738

- Crosier, Danielle M.; Molloy, Daniel P.; Bartholomew, Jerri. "Whirling Disease – Myxobolus cerebralis"z(PDF). Archived from the original (PDF) on 16 February 2008. Retrieved 13 December 2007

- Helfield, J. & Naiman, R. (2006). "Keystone Interactions: Salmon and Bear in Riparian Forests of Alaska" (PDF). Ecosystems. 9 (2): 167–180. doi:10.1007/s10021-004-0063-5

- Meyer CG; Holland KN; Papastamatiou YP (2005). "Sharks can detect changes in the geomagnetic field". Journal of the Royal Society, Interface. 2 (2): 129–30. PMC 1578252. PMID 16849172. doi:10.1098/rsif.2004.0021

- Lichatowich, Jim (1999). Salmon Without Rivers: A History of the Pacific Salmon Crisis. Island Press. ISBN 1559633603. OCLC 868995261

- Guilford, Gwynn (12 March 2015). "Here's why your farmed salmon has color added to it". Quartz (publication). Retrieved 12 March 2015

- Willson MF & Halupka KC (1995). "Anadromous Fish as Keystone Species in Vertebrate Communities" (PDF). Conservation Biology. 9 (3): 489–497. JSTOR 2386604. doi:10.1046/j.1523-1739.1995.09030489.x

- Quinn, T.; Carlson, S.; Gende, S. & Rich, H. (2009). "Transportation of Pacific Salmon Carcasses from Streams to Riparian Forests by Bears" (PDF). Canadian Journal of Zoology. 87 (3): 195–203. doi:10.1139/Z09-004

- E., Taylor, Joseph (2001). Making Salmon: An Environmental History of the Northwest Fisheries Crisis. Univ Of Washington Press. ISBN 0295981148. OCLC 228275619

- "SHARKS & RAYS, SeaWorld/Busch Gardens ANIMALS, CIRCULATORY SYSTEM". Busch Entertainment Corporation. Retrieved 2009-09-03

A Comprehensive Study of Fish Anatomy

Fish anatomy studies the form and structure of fishes which include the body, skeleton, external organs like gills, skin and scales and internal organs like stomach, kidneys, liver and heart of the fish. The chapter on fish anatomy offers an insightful focus, keeping in mind the complex subject matter.

Fish Anatomy

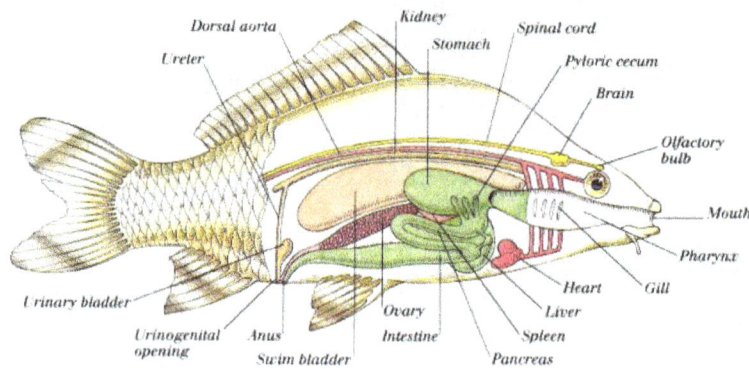

Internal anatomy of a bony fish

Fish anatomy is the study of the form or morphology of fishes. It can be contrasted with fish physiology, which is the study of how the component parts of fish function together in the living fish. In practice, fish anatomy and fish physiology complement each other, the former dealing with the structure of a fish, its organs or component parts and how they are put together, such as might be observed on the dissecting table or under the microscope, and the latter dealing with how those components function together in living fish.

The anatomy of fish is often shaped by the physical characteristics of water, the medium in which fish live. Water is much denser than air, holds a relatively small amount of dissolved oxygen, and absorbs more light than air does. The body of a fish is divided into a head, trunk and tail, although the divisions between the three are not always externally visible. The skeleton, which forms the support structure inside the fish, is either made of cartilage, in cartilaginous fish, or bone in bony fish. The main skeletal element is the vertebral column, composed of articulating vertebrae which are lightweight yet strong. The ribs attach to the spine and there are no limbs or limb girdles. The main external features of the fish, the fins, are composed of either bony or soft spines called rays which, with the exception of the caudal fins, have no direct connection with the spine. They are supported by the muscles which compose the main part of the trunk. The heart has two chambers and pumps the blood through the respiratory surfaces of the gills and on round the body in a single circulatory loop. The eyes are adapted for seeing underwater and have only local vision. There is an inner ear but no external or middle ear. Low frequency vibrations are detected by the lateral line system of sense organs that run along the length of the sides of fish, and these respond to nearby movements and to changes in water pressure.

Sharks and rays are basal fish with numerous primitive anatomical features similar to those of ancient fish, including skeletons composed of cartilage. Their bodies tend to be dorso-ventrally flattened, they usually have five pairs of gill slits and a large mouth set on the underside of the head. The dermis is covered with separate dermal placoid scales. They have a cloaca into which the urinary and genital passages open, but not a swim bladder. Cartilaginous fish produce a small number of large, yolky eggs. Some species are ovoviviparous and the young develop internally but others are oviparous and the larvae develop externally in egg cases.

The bony fish lineage shows more derived anatomical traits, often with major evolutionary changes from the features of ancient fish. They have a bony skeleton, are generally laterally flattened, have five pairs of gills protected by an operculum, and a mouth at or near the tip of the snout. The dermis is covered with overlapping scales. Bony fish have a swim bladder which helps them maintain a constant depth in the water column, but not a cloaca. They mostly spawn a large number of small eggs with little yolk which they broadcast into the water column.

Body

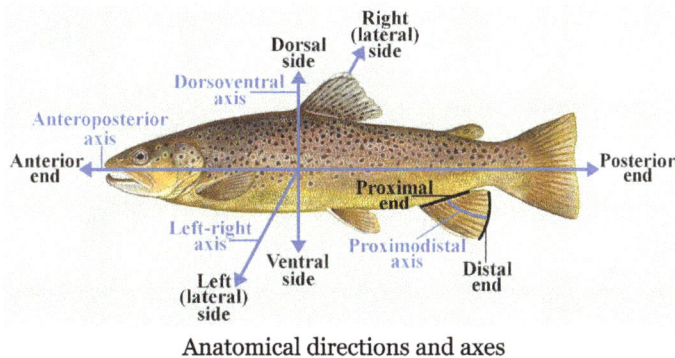

Anatomical directions and axes

In many respects fish anatomy is different from humans and mammals, yet it shares the same basic vertebrate body plan from which all vertebrates have evolved: a notochord, rudimentary vertebrae, and a well-defined head and tail.

Fish have a variety of different body plans. At the broadest level their body is divided into head, trunk, and tail, although the divisions are not always externally visible. The body is often fusiform, a streamlined body plan often found in fast-moving fish. They may also be filiform (eel-shaped) or vermiform (worm-shaped). Also, fish are often either compressed (laterally thin) or depressed (dorso-ventrally flat).

Skeleton

There are two different skeletal types: the exoskeleton, which is the stable outer shell of an organism, and the endoskeleton, which forms the support structure inside the body. The skeleton of the fish is either made of cartilage (cartilaginous fishes) or bones (bony fishes). The main features of the fish, the fins, are bony fin rays and, except for the caudal fin, have no direct connection with the spine. They are supported only by the muscles. The ribs attach to the spine.

Skeleton of a bony fish

Bones are rigid organs that form part of the endoskeleton of vertebrates. They function to move, support, and protect the various organs of the body, produce red and white blood cells and store minerals. Bone tissue is a type of dense connective tissue. Because bones come in a variety of shapes and have a complex internal and external structure they are lightweight, yet strong and hard, in addition to fulfilling their many other functions.

Fish bones have been used to bioremediate lead from contaminated soil.

Vertebrae

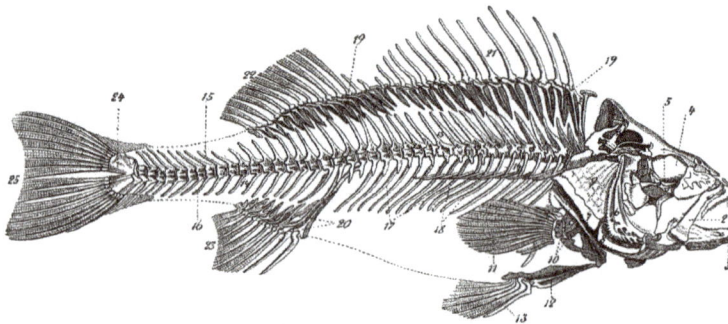

Skeletal structure of a bass showing the vertebral column running from the head to the tail

Skeletal structure of an Atlantic cod

Fish are vertebrates. All vertebrates are built along the basic chordate body plan: a stiff rod running through the length of the animal (vertebral column or notochord), with a hollow tube of nervous tissue (the spinal cord) above it and the gastrointestinal tract below. In all vertebrates, the mouth is found at, or right below, the anterior end of the animal, while the anus opens to the exterior before the end of the body. The remaining part of the body continuing aft of the anus forms a tail with vertebrae and spinal cord, but no gut.

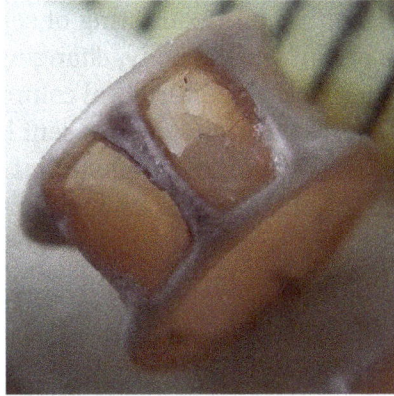

A vertebra (diameter 5 mm) of a small ray-finned fish

The defining characteristic of a vertebrate is the vertebral column, in which the notochord (a stiff rod of uniform composition) found in all chordates has been replaced by a segmented series of stiffer elements (vertebrae) separated by mobile joints (intervertebral discs, derived embryonically and evolutionarily from the notochord). However, a few fish have secondarily lost this anatomy, retaining the notochord into adulthood, such as the sturgeon.

The vertebral column consists of a centrum (the central body or spine of the vertebra), vertebral arches which protrude from the top and bottom of the centrum, and various processes which project from the centrum or arches. An arch extending from the top of the centrum is called a neural arch, while the hemal arch or chevron is found underneath the centrum in the caudal (tail) vertebrae of fish. The centrum of a fish is usually concave at each end (amphicoelous), which limits the motion of the fish. This can be contrasted with the centrum of a mammal, which is flat at each end (acoelous), shaped in a manner that can support and distribute compressive forces.

The vertebrae of lobe-finned fishes consist of three discrete bony elements. The vertebral arch surrounds the spinal cord, and is of broadly similar form to that found in most other vertebrates. Just beneath the arch lies a small plate-like *pleurocentrum*, which protects the upper surface of the notochord, and below that, a larger arch-shaped *intercentrum* to protect the lower border. Both of these structures are embedded within a single cylindrical mass of cartilage. A similar arrangement was found in primitive tetrapods, but, in the evolutionary line that led to reptiles (and hence, also to mammals and birds), the intercentrum became partially or wholly replaced by an enlarged pleurocentrum, which in turn became the bony vertebral body.

In most ray-finned fishes, including all teleosts, these two structures are fused with, and embedded within, a solid piece of bone superficially resembling the vertebral body of mammals. In living amphibians, there is simply a cylindrical piece of bone below the vertebral arch, with no trace of the separate elements present in the early tetrapods.

In cartilagenous fish, such as sharks, the vertebrae consist of two cartilagenous tubes. The upper tube is formed from the vertebral arches, but also includes additional cartilagenous structures filling in the gaps between the vertebrae, and so enclosing the spinal cord in an essentially continuous sheath. The lower tube surrounds the notochord, and has a complex structure, often including multiple layers of calcification.

Lampreys have vertebral arches, but nothing resembling the vertebral bodies found in all higher vertebrates. Even the arches are discontinuous, consisting of separate pieces of arch-shaped cartilage around the spinal cord in most parts of the body, changing to long strips of cartilage above and below in the tail region. Hagfishes lack a true vertebral column, and are therefore not properly considered vertebrates, but a few tiny neural arches are present in the tail. Hagfishes do, however, possess a cranium. For this reason, the vertebrate subphylum is sometimes referred to as "Craniata" when discussing morphology. Molecular analysis since 1992 has suggested that the hagfishes are most closely related to lampreys, and so also are vertebrates in a monophyletic sense. Others consider them a sister group of vertebrates in the common taxon of Craniata.

Head

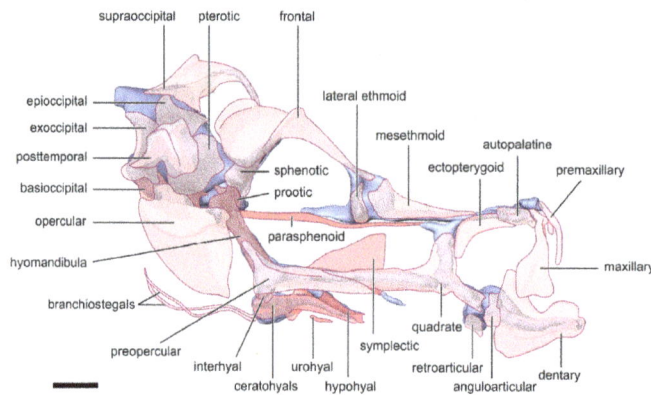

Skull bones as they appear in a seahorse

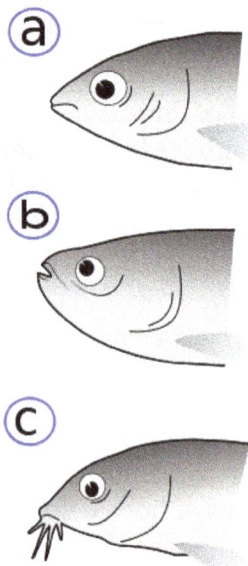

Positions of fish mouths:
(a) - terminal, (b) - superior, (c) - subterminal, inferior

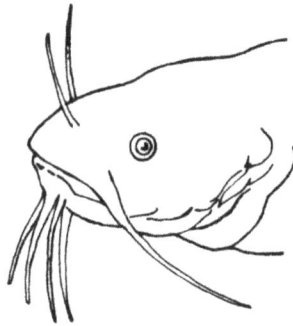

Barbels

The head or skull includes the skull roof (a set of bones covering the brain, eyes and nostrils), the snout (from the eye to the forward most point of the upper jaw), the operculum or gill cover (absent in sharks and jawless fish), and the cheek, which extends from the eye to preopercle. The operculum and preopercle may or may not have spines. In sharks and some primitive bony fish a spiracle, small extra gill opening, is found behind each eye.

The skull in fishes is formed from a series of only loosely connected bones. Jawless fish and sharks only possess a cartilaginous endocranium, with the upper and lower jaws of cartilaginous fish being separate elements not attached to the skull.Bony fishes have additional dermal bone, forming a more or less coherent skull roof in lungfish and holost fish. The lower jaw defines a chin.

In lampreys, the mouth is formed into an oral disk. In most jawed fish, however, there are three general configurations. The mouth may be on the forward end of the head (*terminal*), may be upturned (*superior*), or may be turned downwards or on the bottom of the fish (*subterminal* or *inferior*). The mouth may be modified into a suckermouth adapted for clinging onto objects in fast-moving water.

The simpler structure is found in jawless fish, in which the cranium is represented by a trough-like basket of cartilaginous elements only partially enclosing the brain, and associated with the capsules for the inner ears and the single nostril. Distinctively, these fish have no jaws.

Cartilaginous fish, such as sharks, also have simple, and presumably primitive, skull structures. The cranium is a single structure forming a case around the brain, enclosing the lower surface and the sides, but always at least partially open at the top as a large fontanelle. The most anterior part of the cranium includes a forward plate of cartilage, the rostrum, and capsules to enclose the olfactory organs. Behind these are the orbits, and then an additional pair of capsules enclosing the structure of the inner ear. Finally, the skull tapers towards the rear, where the foramen magnum lies immediately above a single condyle, articulating with the first vertebra. There are, in addition, at various points throughout the cranium, smaller foramina for the cranial nerves. The jaws consist of separate hoops of cartilage, almost always distinct from the cranium proper.

In the ray-finned fishes, there has also been considerable modification from the primitive pattern. The roof of the skull is generally well formed, and although the exact relationship of its bones to those of tetrapods is unclear, they are usually given similar names for convenience. Other elements of the skull, however, may be reduced; there is little cheek region behind the enlarged orbits, and little, if any bone in between them. The upper jaw is often formed largely from the premaxilla, with the maxilla itself located further back, and an additional bone, the symplectic, linking the jaw to the rest of the cranium.

Although the skulls of fossil lobe-finned fish resemble those of the early tetrapods, the same cannot be said of those of the living lungfishes. The skull roof is not fully formed, and consists of multiple, somewhat irregularly shaped bones with no direct relationship to those of tetrapods. The upper jaw is formed from the pterygoids and vomers alone, all of which bear teeth. Much of the skull is formed from cartilage, and its overall structure is reduced.

The head may have several fleshy structures known as barbels, which may be very long and resemble whiskers. Many fish species also have a variety of protrusions or spines on the head. The nostrils or nares of almost all fishes do not connect to the oral cavity, but are pits of varying shape and depth.

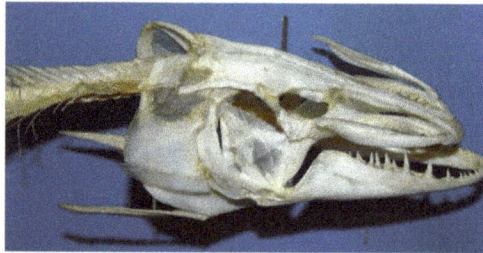

Skull of a northern pike

Skull of *Tiktaalik*, a genus of extinct sarcopterygian (lobe-finned "fish") from the late Devonian period

External Organs

Jaw

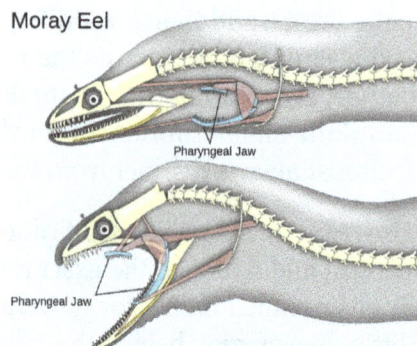

Moray eels have two sets of jaws: the oral jaws that capture prey and the pharyngeal jaws that advance into the mouth and move prey from the oral jaws to the esophagus for swallowing

The vertebrate jaw probably originally evolved in the Silurian period and appeared in the Placoderm fish which further diversified in the Devonian. Jaws are thought to derive from the pharyngeal arches that support the gills in fish. The two most anterior of these arches are thought to have become the jaw itself and the hyoid arch, which braces the jaw against the braincase and increases mechanical efficiency. While there is no fossil evidence directly to support this theory, it makes sense in light of the numbers of pharyngeal arches that are visible in extant jawed (the Gnathostomes), which have seven arches, and primitive jawless vertebrates (the Agnatha), which have nine.

Jaws of great white shark

It is thought that the original selective advantage garnered by the jaw was not related to feeding, but to increased respiration efficiency. The jaws were used in the buccal pump (observable in modern fish and amphibians) that pumps water across the gills of fish or air into the lungs in the case of amphibians. Over evolutionary time the more familiar use of jaws (to humans), in feeding, was selected for and became a very important function in vertebrates.

Linkage systems are widely distributed in animals. The most thorough overview of the different types of linkages in animals has been provided by M. Muller, who also designed a new classification system, which is especially well suited for biological systems. Linkage mechanisms are especially frequent and manifold in the head of bony fishes, such as wrasses, which have evolved many specialized feeding mechanisms. Especially advanced are the linkage mechanisms of jaw protrusion. For suction feeding a system of linked four-bar linkages is responsible for the coordinated opening of the mouth and 3-D expansion of the buccal cavity. Other linkages are responsible for protrusion of the premaxilla.

Eyes

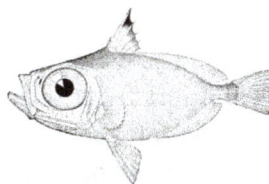

Zenion hololepis is a small deep water fish with large eyes

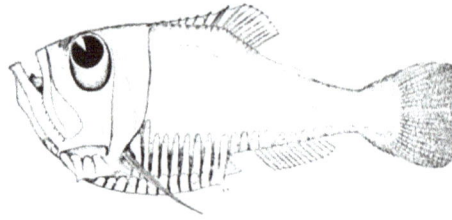

The deep sea half-naked hatchetfish has eyes which look overhead where it can see the silhouettes of prey

Fish eyes are similar to terrestrial vertebrates like birds and mammals, but have a more spherical lens. Their retinas generally have both rod cells and cone cells (for scotopic and photopic vision), and most species have colour vision. Some fish can see ultraviolet and some can see polarized light. Amongst jawless fish, the lamprey has well-developed eyes, while the hagfish has only primitive eyespots. The ancestors of modern hagfish, thought to be the protovertebrate were evidently pushed to very deep, dark waters, where they were less vulnerable to sighted predators, and where it is advantageous to have a convex eye-spot, which gathers more light than a flat or concave one. Unlike humans, fish normally adjust focus by moving the lens closer to or further from the retina.

Gills

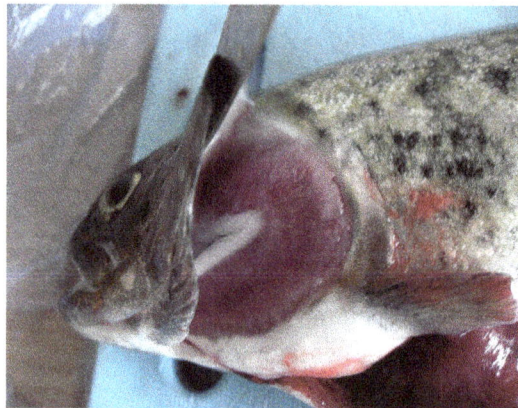

Gill of a rainbow trout

The gills, located under the operculum, are a respiratory organ for the extraction of oxygen from water and for the excretion of carbon dioxide. They are not usually visible, but can be seen in some species, such as the frilled shark. The labyrinth organ of Anabantoidei and Clariidae is used to allow the fish to extract oxygen from the air. Gill rakers are bony or cartilaginous, finger-like projections off the gill arch which function in filter-feeders to retain filtered prey.

Skin

The epidermis of fish consists entirely of live cells, with only minimal quantities of keratin in the cells of the superficial layer. It is generally permeable. The dermis of bony fish typically contains relatively little of the connective tissue found in tetrapods. Instead, in most species, it is largely replaced by solid, protective bony scales. Apart from some particularly large dermal bones that form parts of the skull, these scales are lost in tetrapods, although many reptiles do have scales of a different kind, as do pangolins. Cartilaginous fish have numerous tooth-like denticles embedded in their skin, in place of true scales.

Sweat glands and sebaceous glands are both unique to mammals, but other types of skin glands are found in fish. Fish typically have numerous individual mucus-secreting skin cells that aid in insulation and protection, but may also have poison glands, photophores, or cells that produce a more watery, serous fluid. Melanin colours the skin of many species, but in fish the epidermis is often relatively colourless. Instead, the colour of the skin is largely due to chromatophores in the dermis, which, in addition to melanin, may contain guanine or carotenoid pigments. Many species, such as flounders, change the colour of their skin by adjusting the relative size of their chromatophores.

Scales

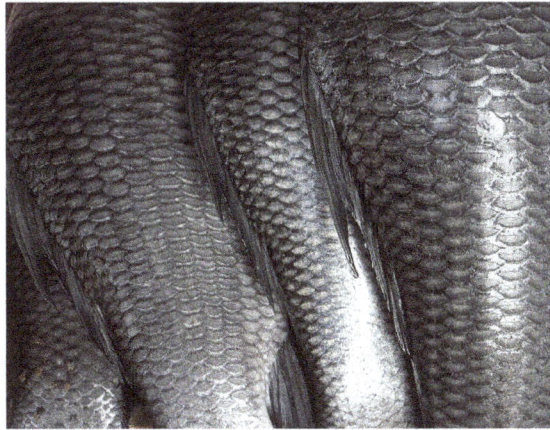

Cycloid scales covering rohu

The outer body of many fish is covered with scales, which are part of the fish's integumentary system. The scales originate from the mesoderm (skin), and may be similar in structure to teeth. Some species are covered instead by scutes. Others have no outer covering on the skin. Most fish are covered in a protective layer of slime (mucus).

There are four principal types of fish scales.

1. Placoid scales, also called dermal denticles, are similar to teeth in that they are made of dentin covered by enamel. They are typical of sharks and rays.

2. Ganoid scales are flat, basal-looking scales that cover a fish body with little overlapping. They are typical of gar and bichirs.

3. Cycloid scales are small oval-shaped scales with growth rings. Bowfin and remora have cycloid scales.

4. Ctenoid scales are similar to the cycloid scales, with growth rings. They are distinguished by spines that cover one edge. Halibut have this type of scale.

Another, less common, type of scale is the scute, which is:

-   an external shield-like bony plate, or

-   a modified, thickened scale that often is keeled or spiny, or

-   a projecting, modified (rough and strongly ridged) scale, usually associated with the lateral line, or on the caudal peduncle forming caudal keels, or along the ventral profile. Some fish, such as pineconefish, are completely or partially covered in scutes.

Lateral Line

The lateral line is clearly visible as a line of receptors running along the side of this Atlantic cod

The lateral line is a sense organ used to detect movement and vibration in the surrounding water. For example, fish can use their lateral line system to follow the vortices produced by fleeing prey. In most species, it consists of a line of receptors running along each side of the fish.

Photophores

Photophores are light-emitting organs which appears as luminous spots on some fishes. The light can be produced from compounds during the digestion of prey, from specialized mitochondrial cells in the organism called photocytes, or associated with symbiotic bacteria, and are used for attracting food or confusing predators.

Fins

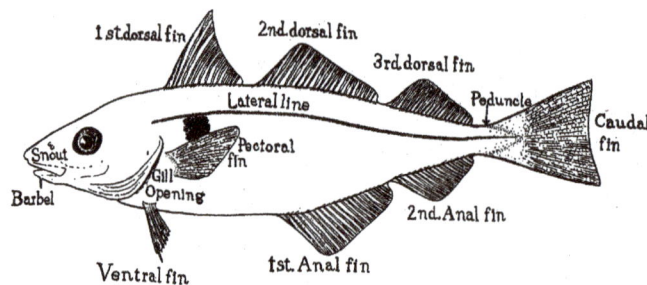

The haddock, a type of cod, is ray-finned. It has three dorsal and two anal fins

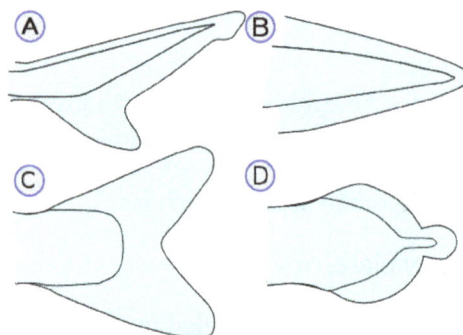

Types of caudal (tail) fin:
(A) - Heterocercal, (B) - Protocercal, (C) - Homocercal, (D) - Diphycercal

Sharks possess a heterocercal caudal fin. The dorsal portion is usually larger than the ventral portion

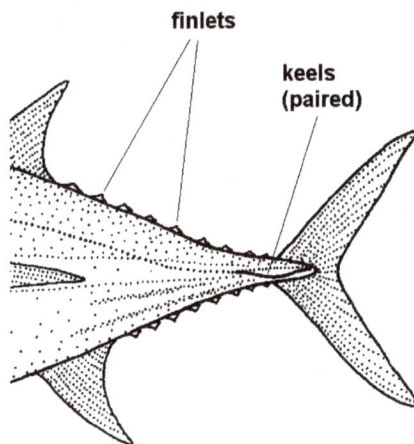

finlets

**keels
(paired)**

The high performance bigeye tuna is equipped with a homocercal
caudal fin and finlets and keels. Drawn by Dr Tony Ayling

Fins are the most distinctive features of fish. They are either composed of bony spines or rays protruding from the body with skin covering them and joining them together, either in a webbed fashion as seen in most bony fish or similar to a flipper as seen in sharks. Apart from the tail or caudal fin, fins have no direct connection with the spine and are supported by muscles only. Their principal function is to help the fish swim. Fins can also be used for gliding or crawling, as seen in the flying fish and frogfish. Fins located in different places on the fish serve different purposes, such as moving forward, turning, and keeping an upright position. For every fin, there are a number of fish species in which this particular fin has been lost during evolution.

Spines and Rays

In bony fish, most fins may have spines or rays. A fin may contain only spiny rays, only soft rays, or a combination of both. If both are present, the spiny rays are always anterior. Spines are generally stiff, sharp and unsegmented. Rays are generally soft, flexible, segmented, and may be branched. This segmentation of rays is the main difference that distinguishes them from spines; spines may be flexible in certain species, but never segmented.

Spines have a variety of uses. In catfish, they are used as a form of defense; many catfish have the ability to lock their spines outwards. Triggerfish also use spines to lock themselves in crevices to prevent them being pulled out.

Lepidotrichia are bony, bilaterally-paired, segmented fin rays found in bony fishes. They develop around actinotrichia as part of the dermal exoskeleton. Lepidotrichia may have some cartilage or bone in them as well. They are actually segmented and appear as a series of disks stacked one on top of another. The genetic basis for the formation of the fin rays is thought to be genes coding for the proteins actinodin 1 and actinodin 2.

Types of Fin

- Dorsal fins are located on the back. Most fishes have one dorsal fin, but some fishes have two or three. The dorsal fins serve to protect the fish against rolling, and assists in sudden turns and stops. In anglerfish, the anterior of the dorsal fin is modified into an *illicium* and *esca*, a biological equivalent to a fishing rod and lure. The bones that support the dorsal fin are called *Pterygiophore*. There are two to three of them: "proximal", "middle", and "distal". In spinous fins the distal is often fused to the middle, or not present at all.

- The caudal fin is the tail fin, located at the end of the caudal peduncle and is used for propulsion. The caudal peduncle is the narrow part of the fish's body to which the caudal or tail fin is attached. The hypural joint is the joint between the caudal fin and the last of the vertebrae. The hypural is often fan-shaped. The tail is called:

  o *Heterocercal* if the vertebrae extend into the upper lobe of the tail, making it longer (as in sharks)

  o *Reversed heterocercal* if the vertebrae extend into the lower lobe of the tail, making it longer (as in the Anaspida)

  o *Protocercal* if the vertebrae extend to the tip of the tail and the tail is symmetrical but not expanded (as in amphioxus)

  o *Diphycercal* if the vertebrae extend to the tip of the tail and the tail is symmetrical and expanded (as in the bichir, lungfish, lamprey and coelacanth. Most Palaeozoic fishes had a diphycercal heterocercal tail.)

  o Most fish have a *homocercal* tail, where the fin appears superficially symmetric but the vertebrae extend for a very short distance into the upper lobe of the fin. This can be expressed in a variety of shapes. The tail fin can be:

    ▪ rounded at the end

    ▪ truncated: or end in a more-or-less vertical edge, such as in salmon

    ▪ forked: or end in two prongs

    ▪ emarginate: or with a slight inward curve.

    ▪ continuous: with dorsal, caudal and anal fins attached, such as in eels

- The anal fin is located on the ventral surface behind the anus. This fin is used to stabilize the fish while swimming.

- The paired pectoral fins are located on each side, usually just behind the operculum, and are homologous to the forelimbs of tetrapods. A peculiar function of pectoral fins, highly developed in some fish, is the creation of the dynamic lifting force that assists some fish, such as sharks, in maintaining depth and also enables the "flight" for flying fish. In many fish, the pectoral fins aid in walking, especially in the lobe-like fins of some anglerfish and in the mudskipper. Certain rays of the pectoral fins may be adapted into finger-like projections, such as in sea robins and flying gurnards. The "horns" of manta rays and their relatives are called *cephalic fins*; this is actually a modification of the anterior portion of the pectoral fin.

- The paired pelvic or ventral fins are located ventrally below the pectoral fins. They are homologous to the hindlimbs of tetrapods. The pelvic fin assists the fish in going up or down through the water, turning sharply, and stopping quickly. In gobies, the pelvic fins are often fused into a single sucker disk. This can be used to attach to objects.

- The adipose fin is a soft, fleshy fin found on the back behind the dorsal fin and just forward of the caudal fin. It is absent in many fish families, but is found in Salmonidae, characins and catfishes. Its function has remained a mystery, and is frequently clipped off to mark hatchery-raised fish, though data from 2005 showed that trout with their adipose fin removed have an 8% higher tailbeat frequency. Additional research published in 2011 has suggested that the fin may be vital for the detection of and response to stimuli such as touch, sound and changes in pressure. Canadian researchers identified a neural network in the fin, indicating that it likely has a sensory function, but are still not sure exactly what the consequences of removing it are.

- Some types of fast-swimming fish have a horizontal caudal keel just forward of the tail fin. Much like the keel of a ship, this is a lateral ridge on the caudal peduncle, usually composed of scutes, that provides stability and support to the caudal fin. There may be a single paired keel, one on each side, or two pairs above and below.

- Finlets are small fins, generally between the dorsal and the caudal fins also between the anal fin and the caudal fin (in bichirs, there are only finlets on the dorsal surface and no dorsal fin). In some fish such as tuna or sauries, they are rayless, non-retractable, and found between the last dorsal and/or anal fin and the caudal fin.

Internal Organs

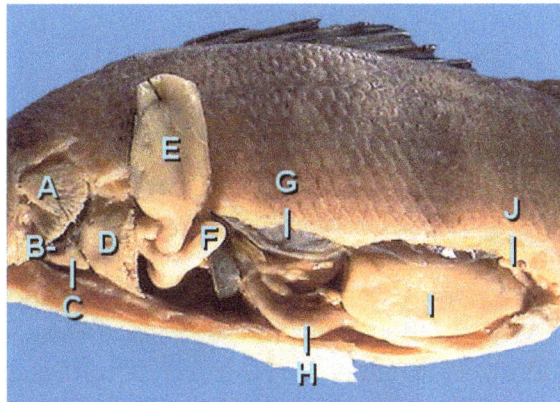

Internal organs of a male yellow perch
A = gill, B = heart atrium, C: heart ventricle, D: liver (cut),
E: stomach, F = pyloric caeca, G = swim bladder, H = intestine, I = testis, J = urinary bladder

Intestines

As with other vertebrates, the intestines of fish consist of two segments, the small intestine and the large intestine. In most higher vertebrates, the small intestine is further divided into the duodenum and other parts. In fish, the divisions of the small intestine are not as clear, and the terms *anterior intestine* or *proximal intestine* may be used instead of duodenum. In bony fish, the intestine is relatively short, typically around one and a half times the length of the fish's body. It commonly has a number of *pyloric caeca*, small pouch-like structures along its length that help to increase the overall surface area of the organ for digesting food. There is no ileocaecal valve in teleosts, with the boundary between the small intestine and the rectum being marked only by the end of the digestive epithelium. There is no small intestine as such in non-teleost fish, such as sharks, sturgeons, and lungfish. Instead, the digestive part of the gut forms a *spiral intestine*, connecting the stomach to the rectum. In this type of gut, the intestine itself is relatively straight, but has a long fold running along the inner surface in a spiral fashion, sometimes for dozens of turns. This fold creates a valve-like structure that greatly increases both the surface area and the effective length of the intestine. The lining of the spiral intestine is similar to that of the small intestine in teleosts and non-mammalian tetrapods. In lampreys, the spiral valve is extremely small, possibly because their diet requires little digestion. Hagfish have no spiral valve at all, with digestion occurring for almost the entire length of the intestine, which is not subdivided into different regions.

Pyloric Caeca

The pyloric caecum is a pouch, usually peritoneal, at the beginning of the large intestine. It receives faecal material from the ileum, and connects to the ascending colon of the large intestine. It is present in most amniotes, and also in lungfish. Many fish in addition have a number of small outpocketings, also called pyloric caeca, along their intestine; despite the name they are not homologous with the caecum of amniotes, and their purpose is to increase the overall area of the digestive epithelium.

The black swallower is a species of deep sea fish with an extensible
stomach which allows it to swallow fish larger than itself

Stomach

As with other vertebrates, the relative positions of the esophageal and duodenal openings to the stomach remain relatively constant. As a result, the stomach always curves somewhat to the left before curving back to meet the pyloric sphincter. However, lampreys, hagfishes, chimaeras, lungfishes, and some teleost fish have no stomach at all, with the esophagus opening directly into the intestine. These fish consume diets that either require little storage of food, or no pre-digestion with gastric juices, or both.

Kidneys

The kidneys of fish are typically narrow, elongated organs, occupying a significant portion of the trunk. They are similar to the mesonephros of higher vertebrates (reptiles, birds and mammals). The kidneys contain clusters of nephrons, serviced by collecting ducts which usually drain into a mesonephric duct. However, the situation is not always so simple. In cartilaginous fish there is also a shorter duct which drains the posterior (metanephric) parts of the kidney, and joins with the mesonephric duct at the bladder or cloaca. Indeed, in many cartilaginous fish, the anterior portion of the kidney may degenerate or cease to function altogether in the adult. Hagfish and lamprey kidneys are unusually simple. They consist of a row of nephrons, each emptying directly into the mesonephric duct.

Spleen

The spleen is found in nearly all vertebrates. It is a non-vital organ, similar in structure to a large lymph node. It acts primarily as a blood filter, and plays important roles in regard to red blood cells and the immune system. In cartilaginous and bony fish it consists primarily of red pulp and is normally a somewhat elongated organ as it actually lies inside the serosal lining of the intestine. The only vertebrates lacking a spleen are the lampreys and hagfishes. Even in these animals, there is a diffuse layer of haematopoeitic tissue within the gut wall, which has a similar structure to red pulp, and is presumed to be homologous with the spleen of higher vertebrates.

Liver

The liver is a large vital organ present in all fish. It has a wide range of functions, including detoxification, protein synthesis, and production of biochemicals necessary for digestion.

Heart

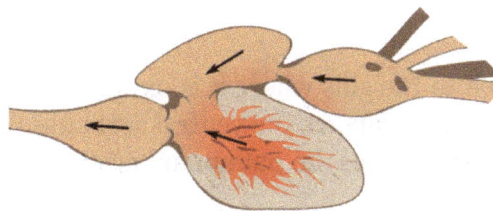

Blood flow through the heart: sinus venosus, atrium, ventricle, and outflow tract

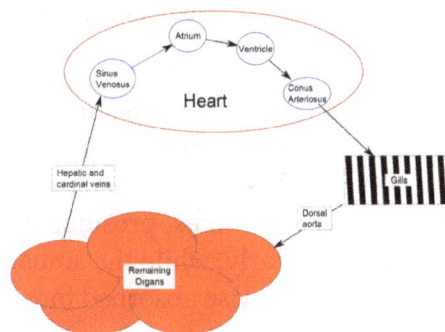

Cardiovascular cycle in a fish

Fish have what is often described as a two-chambered heart, consisting of one atrium to receive blood and one ventricle to pump it, in contrast to three chambers (two atria, one ventricle) of amphibian and most reptile hearts and four chambers (two atria, two ventricles) of mammal and bird hearts. However, the fish heart has entry and exit compartments that may be called chambers, so it is also sometimes described as three-chambered or four-chambered, depending on what is counted as a chamber. The atrium and ventricle are sometimes considered "true chambers", while the others are considered "accessory chambers".

The four compartments are arranged sequentially:

- Sinus venosus, a thin-walled sac or reservoir with some cardiac muscle that collects deoxygenated blood through the incoming hepatic and cardinal veins.

- Atrium, a thicker-walled, muscular chamber that sends blood to the ventricle.

- Ventricle, a thick-walled, muscular chamber that pumps the blood to the fourth part, the outflow tract. The shape of the ventricle varies considerably, usually tubular in fish with elongated bodies, pyramidal with a triangular base in others, or sometimes sac-like in some marine fish.

- The outflow tract (OFT) to the ventral aorta, consisting of the tubular conus arteriosus, bulbus arteriosus, or both. The conus arteriosus, typically found in more primitive species of fish, contracts to assist blood flow to the aorta, while the bulbus anteriosus does not.

Ostial valves, consisting of flap-like connective tissues, prevent blood from flowing backward through the compartments. The ostial valve between the sinus venosus and atrium is called the sino-atrial valve, which closes during ventricular contraction. Between the atrium and ventricle is an ostial valve called the atrio-ventricular valve, and between the bulbus arteriosus and ventricle is an ostial valve called the bulbo-ventricular valve. The conus arteriosus has a variable number of semilunar valves.

The ventral aorta delivers blood to the gills where it is oxygenated and flows, through the dorsal aorta, into the rest of the body. (In tetrapods, the ventral aorta has divided in two; one half forms the ascending aorta, while the other forms the pulmonary artery).

The circulatory systems of all vertebrates, are *closed*. Fish have the simplest circulatory system, consisting of only one circuit, with the blood being pumped through the capillaries of the gills and on to the capillaries of the body tissues. This is known as *single cycle* circulation.

In the adult fish, the four compartments are not arranged in a straight row but, instead form an S-shape with the latter two compartments lying above the former two. This relatively simpler pattern is found in cartilaginous fish and in the ray-finned fish. In teleosts, the conus arteriosus is very small and can more accurately be described as part of the aorta rather than of the heart proper. The conus arteriosus is not present in any amniotes, presumably having been absorbed into the ventricles over the course of evolution. Similarly, while the sinus venosus is present as a vestigial structure in some reptiles and birds, it is otherwise absorbed into the right atrium and is no longer distinguishable.

Swim Bladder

The swim bladder (or gas bladder) is an internal organ that contributes to the ability of a fish to control its buoyancy, and thus to stay at the current water depth, ascend, or descend without having to waste energy in swimming. The bladder is found only in the bony fishes. In the more primitive groups like some minnows, bichirs and lungfish, the bladder is open to the esophagus and doubles as a lung. It is often absent in fast swimming fishes such as the tuna and mackerel families. The condition of a bladder open to the esophagus is called physostome, the closed condition physoclist. In the latter, the gas content of the bladder is controlled through a rete mirabilis, a network of blood vessels effecting gas exchange between the bladder and the blood.

Weberian Apparatus

Fishes of the superorder Ostariophysi possess a structure called the Weberian apparatus, a modification which allow them to hear better. This ability which may well explain the marked success of otophysian fishes. The apparatus is made up of a set of bones known as *Weberian ossicles*, a chain of small bones that connect the auditory system to the swim bladder of fishes. The ossicles connect the gas bladder wall with Y-shaped lymph sinus that abuts the lymph-filled transverse canal joining the sacculi of the right and left ears. This allows the transmission of vibrations to the inner ear. A fully functioning Weberian apparatus consists of the swim bladder, the Weberian ossicles, a portion of the anterior vertebral column, and some muscles and ligaments.

Reproductive Organs

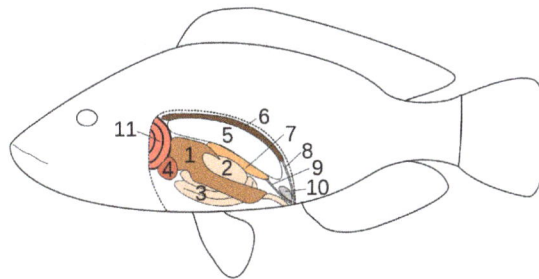

7 = testicles or ovaries

Fish reproductive organs include testes and ovaries. In most species, gonads are paired organs of similar size, which can be partially or totally fused. There may also be a range of secondary organs that increase reproductive fitness. The genital papilla is a small, fleshy tube behind the anus in some fishes, from which the sperm or eggs are released; the sex of a fish often can be determined by the shape of its papilla.

Testes

Most male fish have two testes of similar size. In the case of sharks, the testis on the right side is usually larger. The primitive jawless fish have only a single testis, located in the midline of the body, although even this forms from the fusion of paired structures in the embryo.

Under a tough membranous shell, the tunica albuginea, the testis of some teleost fish, contains very fine coiled tubes called seminiferous tubules. The tubules are lined with a layer of cells (germ cells) that from puberty into old age, develop into sperm cells (also known as spermatozoa or male gametes). The developing sperm travel through the seminiferous tubules to the rete testis located in the mediastinum testis, to the efferent ducts, and then to the epididymis where newly created sperm cells mature. The sperm move into the vas deferens, and are eventually expelled through the urethra and out of the urethral orifice through muscular contractions.

However, most fish do not possess seminiferous tubules. Instead, the sperm are produced in spherical structures called *sperm ampullae*. These are seasonal structures, releasing their contents during the breeding season, and then being reabsorbed by the body. Before the next breeding season, new sperm ampullae begin to form and ripen. The ampullae are otherwise essentially identical to the seminiferous tubules in higher vertebrates, including the same range of cell types.

In terms of spermatogonia distribution, the structure of teleosts testes has two types: in the most common, spermatogonia occur all along the seminiferous tubules, while in Atherinomorph fish they are confined to the distal portion of these structures. Fish can present cystic or semi-cystic spermatogenesis in relation to the release phase of germ cells in cysts to the seminiferous tubules lumen.

Ovaries

Many of the features found in ovaries are common to all vertebrates, including the presence of follicular cells and tunica albuginea There may be hundreds or even millions of fertile eggs present in the ovary of a fish at any given time. Fresh eggs may be developing from the germinal epithelium throughout life. Corpora lutea are found only in mammals, and in some elasmobranch fish; in other species, the remnants of the follicle are quickly resorbed by the ovary. The ovary of teleosts is often contains a hollow, lymph-filled space which opens into the oviduct, and into which the eggs are shed. Most normal female fish have two ovaries. In some elasmobranchs, only the right ovary develops fully. In the primitive jawless fish, and some teleosts, there is only one ovary, formed by the fusion of the paired organs in the embryo.

Fish ovaries may be of three types: gymnovarian, secondary gymnovarian or cystovarian. In the first type, the oocytes are released directly into the coelomic cavity and then enter the ostium, then through the oviduct and are eliminated. Secondary gymnovarian ovaries shed ova into the coelom from which they go directly into the oviduct. In the third type, the oocytes are conveyed to the exterior through the oviduct. Gymnovaries are the primitive condition found in lungfish, sturgeon, and bowfin. Cystovaries characterize most teleosts, where the ovary lumen has continuity with the oviduct. Secondary gymnovaries are found in salmonids and a few other teleosts.

Central Nervous System

Fish typically have quite small brains relative to body size compared with other vertebrates, typically one-fifteenth the brain mass of a similarly sized bird or mammal. However, some fish have relatively large brains, most notably mormyrids and sharks, which have brains about as massive relative to body weight as birds and marsupials.

Fish brains are divided into several regions. At the front are the olfactory lobes, a pair of structures that receive and process signals from the nostrils via the two olfactory nerves. Similar to the way humans smell chemicals in the air, fish smell chemicals in the water by tasting them. The olfactory lobes are very large in fish that hunt primarily by smell, such as hagfish, sharks, and catfish. Behind the olfactory lobes is the two-lobed telencephalon, the structural equivalent to the cerebrum in higher vertebrates. In fish the telencephalon is concerned mostly with olfaction. Together these structures form the forebrain.

The forebrain is connected to the midbrain via the diencephalon (in the diagram, this structure is below the optic lobes and consequently not visible). The diencephalon performs functions associated with hormones and homeostasis. The pineal body lies just above the diencephalon. This structure detects light, maintains circadian rhythms, and controls color changes. The midbrain or mesencephalon contains the two optic lobes. These are very large in species that hunt by sight, such as rainbow trout and cichlids.

The hindbrain or metencephalon is particularly involved in swimming and balance. The cerebellum is a single-lobed structure that is typically the biggest part of the brain. Hagfish and lampreys have relatively small cerebellae, while the mormyrid cerebellum is massive and apparently involved in their electrical sense.

The brain stem or myelencephalon is the brain's posterior. As well as controlling some muscles and body organs, in bony fish at least, the brain stem governs respiration and osmoregulation.

Vertebrates are the only chordate group to exhibit a proper brain. A slight swelling of the anterior end of the dorsal nerve cord is found in the lancelet, though it lacks the eyes and other complex sense organs comparable to those of vertebrates. Other chordates do not show any trends towards cephalisation. The central nervous system is based on a hollow nerve tube running along the length of the animal, from which the peripheral nervous system branches out to innervate the various systems. The front end of the nerve tube is expanded by a thickening of the walls and expansion of the central canal of spinal cord into three primary brain vesicles: The prosencephalon (forebrain), mesencephalon (midbrain) and rhombencephalon (hindbrain), further differentiated in the various vertebrate groups. Two laterally placed eyes form around outgrows from the midbrain, except in hagfish, though this may be a secondary loss. The forebrain is well developed and subdivided in most tetrapods, while the midbrain dominate in many fish and some salamanders. Vesicles of the forebrain are usually paired, giving rise to hemispheres like the cerebral hemispheres in mammals. The resulting anatomy of the central nervous system, with a single, hollow ventral nerve cord topped by a series of (often paired) vesicles is unique to vertebrates.

Cross-section of the brain of a porbeagle shark, with the cerebellum highlighted

Cerebellum

The circuits in the cerebellum are similar across all classes of vertebrates, including fish, reptiles, birds, and mammals. There is also an analogous brain structure in cephalopods with well-developed brains, such as octopuses. This has been taken as evidence that the cerebellum performs functions important to all animal species with a brain.

There is considerable variation in the size and shape of the cerebellum in different vertebrate species. In amphibians, lampreys, and hagfish, the cerebellum is little developed; in the latter two groups, it is barely distinguishable from the brain-stem. Although the spinocerebellum is present in these groups, the primary structures are small paired nuclei corresponding to the vestibulocerebellum.

The cerebellum of cartilaginous and bony fishes is extraordinarily large and complex. In at least one important respect, it differs in internal structure from the mammalian cerebellum: The fish cerebellum does not contain discrete deep cerebellar nuclei. Instead, the primary targets of Purkinje cells are a distinct type of cell distributed across the cerebellar cortex, a type not seen in mammals. In mormyrid fish (a family of weakly electrosensitive freshwater fish), the cerebellum is considerably larger than the rest of the brain put together. The largest part of it is a special structure called the *valvula*, which has an unusually regular architecture and receives much of its input from the electrosensory system.

Most species of fish and amphibians possess a lateral line system that senses pressure waves in water. One of the brain areas that receives primary input from the lateral line organ, the medial octavolateral nucleus, has a cerebellum-like structure, with granule cells and parallel fibers. In electrosensitive fish, the input from the electrosensory system goes to the dorsal octavolateral nucleus, which also has a cerebellum-like structure. In ray-finned fishes (by far the largest group), the optic tectum has a layer — the marginal layer — that is cerebellum-like.

Identified Neurons

A neuron is called *identified* if it has properties that distinguish it from every other neuron in the same animal—properties such as location, neurotransmitter, gene expression pattern, and connectivity—and if every individual organism belonging to the same species has one and only one neuron with the same set of properties. In vertebrate nervous systems very few neurons are "identified" in this sense—in humans, there are believed to be none—but in simpler nervous systems, some or all neurons may be thus unique.

In vertebrates, the best known identified neurons are the gigantic Mauthner cells of fish. Every fish has two Mauthner cells, located in the bottom part of the brainstem, one on the left side and one on the right. Each Mauthner cell has an axon that crosses over, innervating neurons at the same brain level and then travelling down through the spinal cord, making numerous connections as it goes. The synapses generated by a Mauthner cell are so powerful that a single action potential gives rise to a major behavioral response: within milliseconds the fish curves its body into a C-shape, then straightens, thereby propelling itself rapidly forward. Functionally this is a fast escape response, triggered most easily by a strong sound wave or pressure wave impinging on the lateral line organ of the fish. Mauthner cells are not the only identified neurons in fish—there are about 20 more types, including pairs of "Mauthner cell analogs" in each spinal segmental nucleus. Although a Mauthner cell is capable of bringing about an escape response all by itself, in the context of ordinary behavior other types of cells usually contribute to shaping the amplitude and direction of the response.

Mauthner cells have been described as command neurons. A command neuron is a special type of identified neuron, defined as a neuron that is capable of driving a specific behavior all by itself. Such neurons appear most commonly in the fast escape systems of various species—the squid giant axon and squid giant synapse, used for pioneering experiments in neurophysiology because of their enormous size, both participate in the fast escape circuit of the squid. The concept of a command neuron has, however, become controversial, because of studies showing that some neurons that initially appeared to fit the description were really only capable of evoking a response in a limited set of circumstances.

Immune System

Immune organs vary by type of fish. In the jawless fish (lampreys and hagfish), true lymphoid organs are absent. These fish rely on regions of lymphoid tissue within other organs to produce immune cells. For example, erythrocytes, macrophages and plasma cells are produced in the anterior kidney (or pronephros) and some areas of the gut (where granulocytes mature.) They resemble primitive bone marrow in hagfish. Cartilaginous fish (sharks and rays) have a more advanced immune system. They have three specialized organs that are unique to chondrichthyes; the epigonal organs (lymphoid tissue similar to mammalian bone) that surround the gonads, the Leydig's organ within the walls of their esophagus, and a spiral valve in their intestine. These organs house typical immune cells (granulocytes, lymphocytes and plasma cells). They also possess an identifiable thymus and a well-developed spleen (their most important immune organ) where various lymphocytes, plasma cells and macrophages develop and are stored. Chondrostean fish (sturgeons, paddlefish and bichirs) possess a major site for the production of granulocytes within a mass that is associated with the meninges (membranes surrounding the central nervous system.) Their heart is frequently covered with tissue that contains lymphocytes, reticular cells and a small number of macrophages. The chondrostean kidney is an important hemopoietic organ; where erythrocytes, granulocytes, lymphocytes and macrophages develop.

Like chondrostean fish, the major immune tissues of bony fish (or teleostei) include the kidney (especially the anterior kidney), which houses many different immune cells. In addition, teleost fish possess a thymus, spleen and scattered immune areas within mucosal tissues (e.g. in the skin, gills, gut and gonads). Much like the mammalian immune system, teleost erythrocytes, neutrophils and granulocytes are believed to reside in the spleen whereas lymphocytes are the major cell type found in the thymus. In 2006, a lymphatic system similar to that in mammals was described in one species of teleost fish, the zebrafish. Although not confirmed as yet, this system presumably will be where naive (unstimulated) T cells accumulate while waiting to encounter an antigen.

Fish Jaw

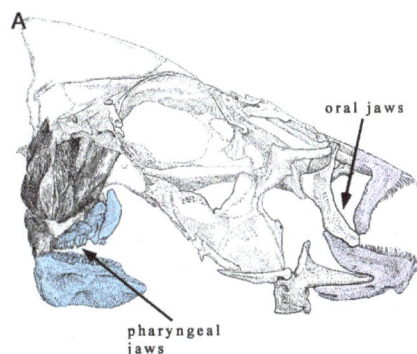

Skull of a generalized cichlid, showing a lateral view of the oral jaws (purple) and the pharyngeal jaws (blue)

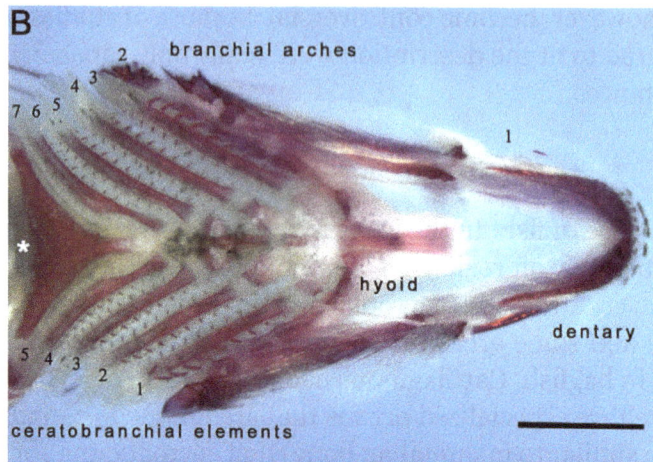

Dorsal view of the lower pharyngeal and oral jaws of a juvenile Malawi eyebiter showing
the branchial (pharyngeal) arches and ceratobranchial elements (arch bones).
The white asterisk indicates the toothed pharyngeal jaw. Scale bar represents 500 μm.

Most bony fishes have two sets of jaws made mainly of bone. The primary oral jaws open and close the mouth, and a second set of pharyngeal jaws are positioned at the back of the throat. The oral jaws are used to capture and manipulate prey by biting and crushing. The pharyngeal jaws, so-called because they are positioned within the pharynx, are used to further process the food and move it from the mouth to the stomach.

Cartilaginous fishes, such as sharks and rays, have one set of oral jaws made mainly of cartilage. They do not have pharyngeal jaws. Generally jaws are articulated and oppose vertically, comprising an upper jaw and a lower jaw and can bear numerous ordered teeth. Bony fishes usually develop only one set of teeth *(monophyodont)*. Cartilaginous fishes grow multiple sets *(polyphyodont)* and replace teeth as they wear.

Jaws probably originated in the pharyngeal arches supporting the gills of jawless fish. The earliest jaws appeared in the now extinct placoderms and spiny sharks during the Silurian, about 430 million years ago. The original selective advantage offered by the jaw was probably not related to feeding, but to increased respiration efficiency — the jaws were used in the buccal pump to pump water across the gills. The familiar use of jaws for feeding would then have developed as a secondary function before becoming the primary function in many vertebrates. All vertebrate jaws, including the human jaw, evolved from early fish jaws. The appearance of the early vertebrate jaw has been described as "perhaps the most profound and radical evolutionary step in the vertebrate history". Fish without jaws had more difficulty surviving than fish with jaws, and most jawless fish became extinct.

Jaws use linkage mechanisms. These linkages can be especially common and complex in the head of bony fishes, such as wrasses, which have evolved many specialized feeding mechanisms. Especially advanced are the linkage mechanisms of jaw protrusion. For suction feeding a system of linked four-bar linkages is responsible for the coordinated opening of the mouth and the three-dimensional expansion of the buccal cavity. Other linkages are responsible for protrusion of the premaxilla. Linkage systems are widely distributed in animals. The most thorough overview of the different types of linkages in animals has been provided by M. Muller, who also designed a new classification system, which is especially well suited for biological systems.

Skull

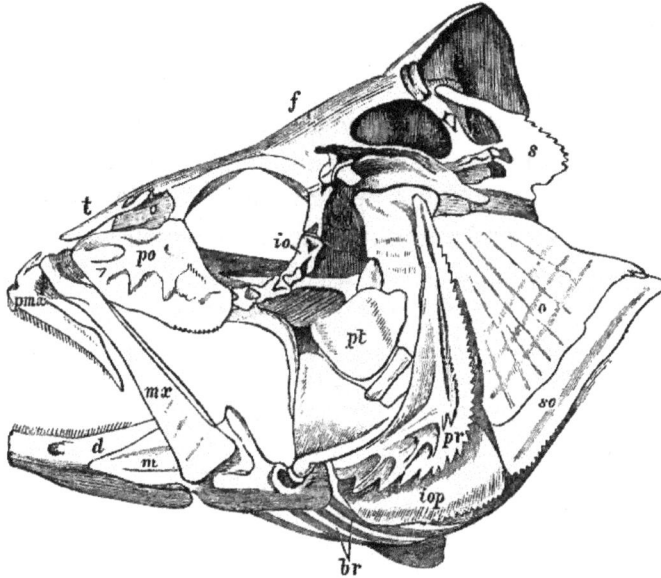

Skeleton of Head of a Perch.

f, frontal.	*pt*, posttympanic.
t, turbinal.	*s*, suprascapula.
po, preorbital.	*o*, opercle.
io, infraorbital ring.	*so*, subopercle.
mx, maxillary.	*pr*, preopercle.
pmx, premaxillary.	*iop*, interopercle.
m, mandible.	*br*, branchiostegal rays.
d, dentary bone.	

The skull of fishes is formed from a series of loosely connected bones. Lampreys and sharks only possess a cartilaginous endocranium, with both the upper and lower jaws being separate elements. Bony fishes have additional dermal bone, forming a more or less coherent skull roof in lungfish and holost fish. The lower jaw defines a chin.

The simpler structure is found in jawless fish, in which the cranium is represented by a trough-like basket of cartilaginous elements only partially enclosing the brain, and associated with the capsules for the inner ears and the single nostril. Distinctively, these fish have no jaws.

Cartilaginous fish, such as sharks, also have simple skulls. The cranium is a single structure forming a case around the brain, enclosing the lower surface and the sides, but always at least partially open at the top as a large fontanelle. The most anterior part of the cranium includes a forward plate of cartilage, the rostrum, and capsules to enclose the olfactory organs. Behind these are the orbits, and then an additional pair of capsules enclosing the structure of the inner ear. Finally, the skull tapers towards the rear, where the foramen magnum lies immediately above a single condyle, articulating with the first vertebra. There are, in addition, at various points throughout the cranium, smaller foramina for the cranial nerves. The jaws consist of separate hoops of cartilage, almost always distinct from the cranium proper.

In ray-finned fishes, there has also been considerable modification from the primitive pattern. The roof of the skull is generally well formed, and although the exact relationship of its bones to those of tetrapods is unclear, they are usually given similar names for convenience. Other elements of the skull, however, may be reduced; there is little cheek region behind the enlarged orbits, and little, if any bone in between them. The upper jaw is often formed largely from the premaxilla, with the maxilla itself located further back, and an additional bone, the symplectic, linking the jaw to the rest of the cranium.

Although the skulls of fossil lobe-finned fish resemble those of the early tetrapods, the same cannot be said of those of the living lungfishes. The skull roof is not fully formed, and consists of multiple, somewhat irregularly shaped bones with no direct relationship to those of tetrapods. The upper jaw is formed from the pterygoids and vomers alone, all of which bear teeth. Much of the skull is formed from cartilage, and its overall structure is reduced.

Oral Jaws

Lower

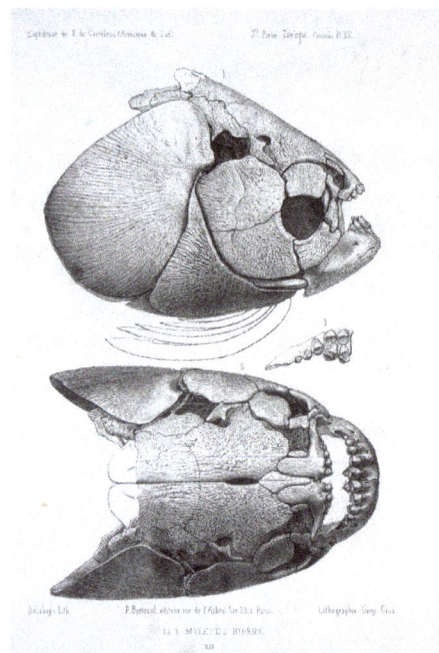

Oral jaw from side and above of *Piaractus brachypomus*, a close relative of piranhas

In vertebrates, the lower jaw (mandible or jawbone) is a bone forming the skull with the cranium. In lobe-finned fishes and the early fossil tetrapods, the bone homologous to the mandible of mammals is merely the largest of several bones in the lower jaw. It is referred to as the *dentary bone*, and forms the body of the outer surface of the jaw. It is bordered below by a number of splenial bones, while the angle of the jaw is formed by a lower angular bone and a suprangular bone just above it. The inner surface of the jaw is lined by a *prearticular* bone, while the articular bone forms the articulation with the skull proper. Finally a set of three narrow *coronoid bones* lie above the prearticular bone. As the name implies, the majority of the teeth are attached to the dentary, but there are commonly also teeth on the coronoid bones, and sometimes on the prearticular as well.

This complex primitive pattern has, however, been simplified to various degrees in the great majority of vertebrates, as bones have either fused or vanished entirely. In teleosts, only the dentary, articular, and angular bones remain. Cartilaginous fish, such as sharks, do not have any of the bones found in the lower jaw of other vertebrates. Instead, their lower jaw is composed of a cartilagenous structure homologous with the Meckel's cartilage of other groups. This also remains a significant element of the jaw in some primitive bony fish, such as sturgeons.

Upper

The upper jaw, or maxilla is a fusion of two bones along the palatal fissure that form the upper jaw. This is similar to the mandible (lower jaw), which is also a fusion of two halves at the mandibular symphysis. In bony fish, the maxilla is called the "upper maxilla," with the mandible being the "lower maxilla". The alveolar process of the maxilla holds the upper teeth, and is referred to as the maxillary arch. In most vertebrates, the foremost part of the upper jaw, to which the incisors are attached in mammals consists of a separate pair of bones, the premaxillae. In bony fish, both maxilla and premaxilla are relatively plate-like bones, forming only the sides of the upper jaw, and part of the face, with the premaxilla also forming the lower boundary of the nostrils. Cartilaginous fish, such as sharks and rays also lack a true maxilla. Their upper jaw is instead formed from a cartilagenous bar that is not homologous with the bone found in other vertebrates.

Some fish have permanently protruding upper jawbones called rostrums. Billfish (marlin, swordfish and sailfish) use rostrums (bills) to slash and stun prey. Paddlefish, goblin sharks and hammerhead sharks have rostrums packed with electroreceptors which signal the presence of prey by detecting weak electrical fields. Sawsharks and the critically endangered sawfish have rostrums (saws) which are both electro-sensitive and used for slashing. The rostrums extend ventrally in front of the fish. In the case of hammerheads the rostrum (hammer) extends both ventrally and laterally (sideways).

Fish with rostrums (extended upper jawbones):

Sailfish, like all billfish, have a rostrum (bill) which evolved from the upper jawbone

The paddlefish has a rostrum packed with electroreceptors

Hammerhead sharks use their rostrum (hammer) to detect and pin rays buried in sand

Pharyngeal Jaws

Pharyngeal jaws are a second set of jaws distinct from the primary (oral) jaws. They are contained within the throat, or pharynx, of most bony fish. They are believed to have originated, in a similar way to oral jaws, as a modification of the fifth gill arch which no longer has a respiratory function. The first four arches still function as gills. Unlike the oral jaw, the pharyngeal jaw has no jaw joint, but is supported instead by a sling of muscles.

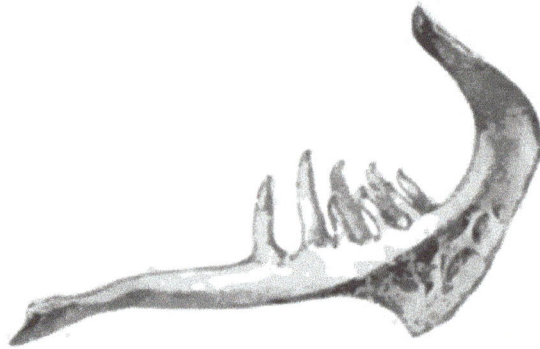

Pharyngeal jaw of an asp carrying some pharyngeal teeth

A notable example occurs with the moray eel. The pharyngeal jaws of most fishes are not mobile. The pharyngeal jaws of the moray are highly mobile, perhaps as an adaptation to the constricted nature of the burrows they inhabit which inhibits their ability to swallow as other fishes do by creating a negative pressure in the mouth. Instead, when the moray bites prey, it first bites normally with its oral jaws, capturing the prey. Immediately thereafter, the pharyngeal jaws are brought forward and bite down on the prey to grip it; they then retract, pulling the prey down the moray eel's gullet, allowing it to be swallowed.

All vertebrates have a pharynx, used in both feeding and respiration. The pharynx arises during development through a series of six or more outpocketings called pharyngeal arches on the lateral sides of the head. The pharyngeal arches give rise to a number of different structures in the skeletal, muscular and circulatory systems in a manner which varies across the vertebrates. Pharyngeal arches trace back through chordates to basal deuterostomes who also share endodermal outpocketings of the pharyngeal apparatus. Similar patterns of gene expression can be detected in the developing pharynx of amphioxus and hemichordates. However, the vertebrate pharynx is unique in that it gives rise to endoskeletal support through the contribution of neural crest cells.

Cartilaginous Jaws

Cartilaginous fishes (sharks, rays and skates) have cartilaginous jaws. The jaw's surface (in comparison to the vertebrae and gill arches) needs extra strength due to its heavy exposure to physical stress. It has a layer of tiny hexagonal plates called "tesserae", which are crystal blocks of calcium salts arranged as a mosaic. This gives these areas much of the same strength found in the bony tissue found in other animals.

Generally sharks have only one layer of tesserae, but the jaws of large specimens, such as the bull shark, tiger shark, and the great white shark, have two to three layers or more, depending on body size. The jaws of a large great white shark may have up to five layers. In the rostrum (snout), the cartilage can be spongy and flexible to absorb the power of impacts.

In sharks and other extant elasmobranchs the upper jaw is not fused to the cranium, and the lower jaw is articulated with the upper. The arrangement of soft tissue and any additional articulations connecting these elements is collectively known as the jaw suspension. There are several archetypal jaw suspensions: amphistyly, orbitostyly, hyostyly, and euhyostyly. In amphistyly, the palatoquadrate has a postorbital articulation with the chondrocranium from which ligaments primarily suspend it anteriorly. The hyoid articulates with the mandibular arch posteriorly, but it appears to provide little support to the upper and lower jaws. In orbitostyly, the orbital process hinges with the orbital wall and the hyoid provides the majority of suspensory support. In contrast, hyostyly involves an ethmoid articulation between the upper jaw and the cranium, while the hyoid most likely provides vastly more jaw support compared to the anterior ligaments. Finally, in euhyostyly, also known as true hyostyly, the mandibular cartilages lack a ligamentous connection to the cranium. Instead, the hyomandibular cartilages provide the only means of jaw support, while the ceratohyal and basihyal elements articulate with the lower jaw, but are disconnected from the rest of the hyoid.

Teeth

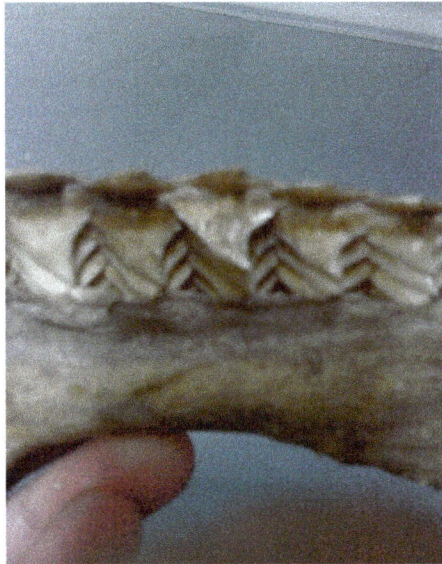

Inside of a shark jaw where new teeth move forward as though on a conveyor belt

Jaws provide a platform in most fishes for simple pointed teeth. Lungfish and chimaera have teeth modified into broad enamel plates with jagged ridges for crushing or grinding. Carp and loach have pharyngeal teeth only. Sea horses, pipefish and adult sturgeon have no teeth of any type.

Shark teeth are embedded in the gums rather than directly affixed to the jaw, and are constantly replaced throughout life. Multiple rows of replacement teeth grow in a groove on the inside of the jaw and steadily moving forward as though on a conveyor belt. Some sharks lose 30,000 or more teeth in their lifetime. The rate of tooth replacement varies from once every 8 to 10 days to several months. In most species, teeth are replaced one at a time as opposed to the simultaneous replacement of an entire row, which is observed in the cookiecutter shark.

Tooth shape depends on the shark's diet: those that feed on mollusks and crustaceans have dense and flattened teeth used for crushing, those that feed on fish have needle-like teeth for gripping, and those that feed on larger prey such as mammals have pointed lower teeth for gripping and triangular upper teeth with serrated edges for cutting. The teeth of plankton-feeders such as the basking shark are small and non-functional.

Cartilaginous jaws and their teeth:

Jaw reconstruction of the extinct *Carcharodon megalodon*, 1909	The thornback ray has teeth adapted to feed on crabs, shrimps and small fish.	The prickly shark has knife-like teeth with main cusps flanked by lateral cusplets

Examples

Salmon

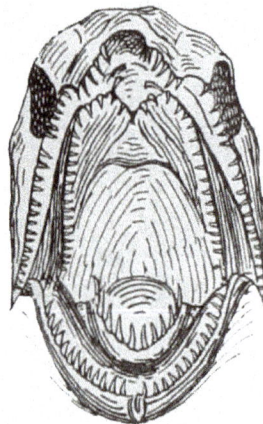

Open mouth of a salmon showing the second set of pharyngeal jaws positioned at the back of the throat

Kype of a spawning male salmon

Male salmon often remodel their jaws during spawning runs so they have a pronounced curvature. These hooked jaws are called kypes. The purpose of the kype is not altogether clear, though they can be used to establish dominance by clamping them around the base of the tail (caudal peduncle) of an opponent.

Cichlids

Fish jaws, like vertebrates in general, normally show bilateral symmetry. An exception occurs with the parasitic scale-eating cichlid *Perissodus microlepis*. The jaws of this fish occur in two distinct morphological forms. One morph has its jaw twisted to the left, allowing it to eat scales more readily on its victim's right flank. The other morph has its jaw twisted to the right, which makes it easier to eat scales on its victim's left flank. The relative abundance of the two morphs in populations is regulated by frequency-dependent selection.

Wrasses

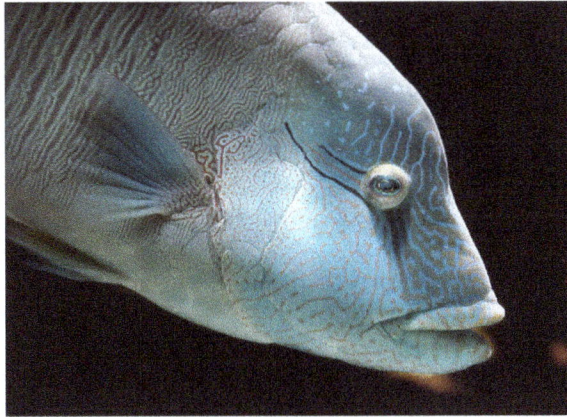

Lips of a humphead wrasse

Wrasses have become a primary study species in fish-feeding biomechanics due to their jaw structure. They have protractile mouths, usually with separate jaw teeth that jut outwards. Many species can be readily recognized by their thick lips, the inside of which is sometimes curiously folded, a peculiarity which gave rise the German name of "lip-fishes" (*Lippfische*.)

The nasal and mandibular bones are connected at their posterior ends to the rigid neurocranium, and the superior and inferior articulations of the maxilla are joined to the anterior tips of these two bones, respectively, creating a loop of 4 rigid bones connected by moving joints. This "four-bar linkage" has the property of allowing numerous arrangements to achieve a given mechanical result (fast jaw protrusion or a forceful bite), thus decoupling morphology from function. The actual morphology of wrasses reflects this, with many lineages displaying different jaw morphology that results in the same functional output in a similar or identical ecological niche.

Other

Relative to its size the stoplight loosejaw has one of the widest gapes of any fish.

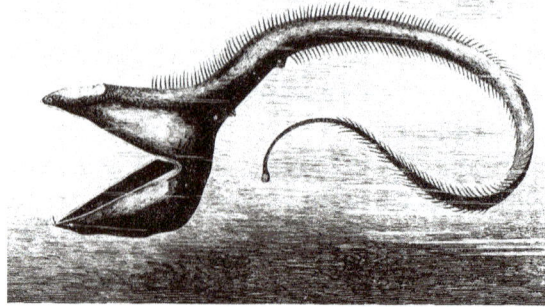

The pelican eel jaws are larger than its body.

Stoplight loosejaws are small fish found worldwide in the deep sea. Relative to their size they have one of the widest gapes of any fish. The lower jaw has no ethmoid membrane (floor) and is attached only by the hinge and a modified tongue bone. There are several large, fang-like teeth in the front of the jaws, followed by many small barbed teeth. There are several groups of pharyngeal teeth that serve to direct food down the esophagus.

Another deep sea fish, the pelican eel, has jaws larger than its body. The jaws are lined with small teeth and are loosely hinged. They open wide enough to swallow a fish larger than the eel itself.

Distichodontidae are a family of fresh water fishes which can be divided into genera with protractile upper jaws which are carnivores, and genera with nonprotractile upper jaws which are herbivores or predators of very small organisms.

Evolution

Vertebrate Classes

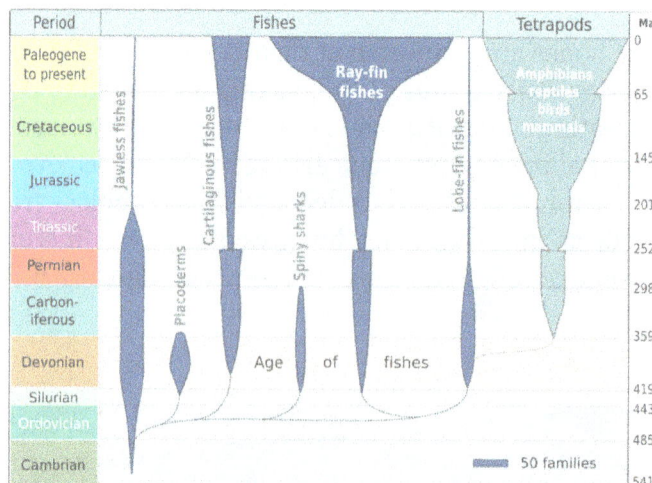

Spindle diagram for the evolution of fish and other vertebrate classes.
The earliest classes that developed jaws were the now extinct placoderms and the spiny sharks.

The appearance of the early vertebrate jaw has been described as "a crucial innovation" and "perhaps the most profound and radical evolutionary step in the vertebrate history". Fish without jaws had more difficulty surviving than fish with jaws, and most jawless fish became extinct. However studies of the cyclostomes, the jawless hagfishes and lampreys that did survive, have yielded little insight into the deep remodelling of the vertebrate skull that must have taken place as early jaws evolved.

The customary view is that jaws are homologous to the gill arches. In jawless fishes a series of gills opened behind the mouth, and these gills became supported by cartilaginous elements. The first set of these elements surrounded the mouth to form the jaw. The upper portion of the second embryonic arch supporting the gill became the hyomandibular bone of jawed fishes, which supports the skull and therefore links the jaw to the cranium. The hyomandibula is a set of bones found in the hyoid region in most fishes. It usually plays a role in suspending the jaws or the operculum in the case of teleosts.

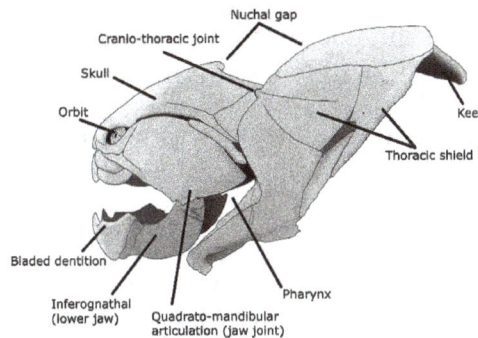

↑ Skull diagram of the huge predatory placoderm fish *Dunkleosteus terrelli*, which lived about 380–360 million years ago

↑ Reconstruction of *Dunkleosteus terrelli*

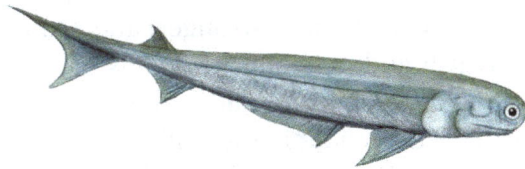

↑ Spiny shark

It is now accepted that the precursors of the jawed vertebrates are the long extinct bony (armoured) jawless fish, the so-called ostracoderms. The earliest known fish with jaws are the now extinct placoderms and spiny sharks.

Placoderms were a class of fish, heavily armoured at the front of their body, which first appeared in the fossil records during the Silurian about 430 million years ago. Initially they very successful, diversifying remarkably during the Devonian. They became extinct by the end of that period, about 360 million years ago. Their largest species, *Dunkleosteus terrelli*, measured up to 10 m (33 ft) and weighed 3.6 t (4.0 short tons). It possessed a four bar linkage mechanism for jaw opening that incorporated connections between the skull, the thoracic shield, the lower jaw and the jaw muscles joined together by movable joints. This mechanism allowed *Dunkleosteus terrelli* to achieve a high speed of jaw opening, opening their jaws in 20 milliseconds and completing the whole process in 50-60 milliseconds, comparable to modern fishes that use suction feeding to assist in prey capture. They could also produce high bite forces when closing the jaw, estimated at 6,000 N (1,350 lb$_f$) at the tip and 7,400 N (1,660 lb$_f$) at the blade edge in the largest individuals. The pressures generated in those regions were high enough to puncture or cut through cuticle or dermal armour suggesting that *Dunkleosteus terrelli* was perfectly adapted to prey on free-swimming, armoured prey like arthropods, ammonites, and other placoderms.

Spiny sharks were another class of fish which appeared also in the fossil records during the Silurian at about the same time as the placoderms. They were smaller than most placoderms, usually under 20 centimetres. Spiny sharks did not diversify as much as placoderms, but survived much longer into the Early Permian about 290 million years ago.

The original selective advantage offered by the jaw was not related to feeding, but to increased respiration efficiency. The jaws were used in the buccal pump still observable in modern fish and amphibians, that uses "breathing with the cheeks" to pump water across the gills of fish or air into the lungs in the case of amphibians. Over evolutionary time the more familiar use of jaws (to humans) for feeding was selected for and became a very important function in vertebrates. Many teleost fish have substantially modified jaws for suction feeding and jaw protrusion, resulting in highly complex jaws with dozens of bones involved.

Jaws are thought to derive from the pharyngeal arches that support the gills in fish. The two most anterior of these arches are thought to have become the jaw itself and the hyoid arch, which braces the jaw against the braincase and increases mechanical efficiency. While there is no fossil evidence directly to support this theory, it makes sense in light of the numbers of pharyngeal arches that are visible in extant jawed (the Gnathostomes), which have seven arches, and primitive jawless vertebrates (the Agnatha), which have nine.

Meckel's cartilage is a piece of cartilage from which the mandibles (lower jaws) of vertebrates evolved. Originally it was the lower of two cartilages which supported the first gill arch (nearest the front) in early fish. Then it grew longer and stronger, and acquired muscles capable of closing the developing jaw. In early fish and in chondrichthyans (cartilaginous fish such as sharks), Meckel's cartilage continued to be the main component of the lower jaw. But in the adult forms of osteichthyans (bony fish) and their descendants (amphibians, reptiles, birds and mammals) the cartilage was covered in bone - although in their embryos the jaw initially develops as the Meckel's cartilage. In tetrapods the cartilage partially ossifies (changes to bone) at the rear end of the jaw and becomes the articular bone, which forms part of the jaw joint in all tetrapods except mammals.

Gill

A gill is a respiratory organ found in many aquatic organisms that extracts dissolved oxygen from water and excretes carbon dioxide. The gills of some species, such as hermit crabs, have adapted to allow respiration on land provided they are kept moist. The microscopic structure of a gill presents a large surface area to the external environment.

The red gills of this common carp are visible as a result of a gill flap birth defect.

Many microscopic aquatic animals, and some larger but inactive ones, can absorb adequate oxygen through the entire surface of their bodies, and so can respire adequately without a gill. However, more complex or more active aquatic organisms usually require a gill or gills.

Gills usually consist of thin filaments of tissue, branches, or slender, tufted processes that have a highly folded surface to increase surface area. A high surface area is crucial to the gas exchange of aquatic organisms, as water contains only a small fraction of the dissolved oxygen that air does. A cubic meter of air contains about 250 grams of oxygen at STP. The concentration of oxygen in water is lower than in air and it diffuses more slowly. In fresh water, the dissolved oxygen content is approximately 8 cm^3/L compared to that of air which is 210 cm^3/L. Water is 777 times more dense than air and is 100 times more viscous. Oxygen has a diffusion rate in air 10,000 times greater than in water. The use of sac-like lungs to remove oxygen from water would not be efficient enough to sustain life. Rather than using lungs, "[g]aseous exchange takes place across the surface of highly vascularised gills over which a one-way current of water is kept flowing by a specialised pumping mechanism. The density of the water prevents the gills from collapsing and lying on top of each other, which is what happens when a fish is taken out of water."

With the exception of some aquatic insects, the filaments and lamellae (folds) contain blood or coelomic fluid, from which gases are exchanged through the thin walls. The blood carries oxygen to other parts of the body. Carbon dioxide passes from the blood through the thin gill tissue into the water. Gills or gill-like organs, located in different parts of the body, are found in various groups of aquatic animals, including mollusks, crustaceans, insects, fish, and amphibians.

Vertebrate Gills

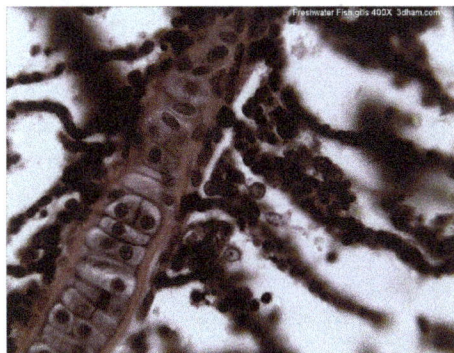

Freshwater Fish Gills magnified 400 times

The gills of vertebrates typically develop in the walls of the pharynx, along a series of gill slits opening to the exterior. Most species employ a countercurrent exchange system to enhance the diffusion of substances in and out of the gill, with blood and water flowing in opposite directions to each other. The gills are composed of comb-like filaments, the gill lamellae, which help increase their surface area for oxygen exchange.

When a fish breathes, it draws in a mouthful of water at regular intervals. Then it draws the sides of its throat together, forcing the water through the gill openings, so it passes over the gills to the outside. Fish gill slits may be the evolutionary ancestors of the tonsils, thymus glands, and Eustachian tubes, as well as many other structures derived from the embryonic branchial pouches.

Fish

Cartilaginous Fish

Sharks and rays typically have five pairs of gill slits that open directly to the outside of the body, though some more primitive sharks have six pairs and the Broadnose sevengill shark being the only cartilaginous fish exceeding this number. Adjacent slits are separated by a cartilaginous gill arch from which projects a cartilaginous gill ray. This gill ray is the support for the sheet-like interbranchial septum, which the individual lamellae of the gills lie on either side of. The base of the arch may also support gill rakers, projections into the pharyngeal cavity that help to prevent large pieces of debris from damaging the delicate gills.

A smaller opening, the spiracle, lies in the back of the first gill slit. This bears a small pseudobranch that resembles a gill in structure, but only receives blood already oxygenated by the true gills. The spiracle is thought to be homologous to the ear opening in higher vertebrates.

Most sharks rely on ram ventilation, forcing water into the mouth and over the gills by rapidly swimming forward. In slow-moving or bottom-dwelling species, especially among skates and rays, the spiracle may be enlarged, and the fish breathes by sucking water through this opening, instead of through the mouth.

Chimaeras differ from other cartilagenous fish, having lost both the spiracle and the fifth gill slit. The remaining slits are covered by an operculum, developed from the septum of the gill arch in front of the first gill.

Bony Fish

The red gills inside a detached tuna head (viewed from behind)

In bony fish, the gills lie in a branchial chamber covered by a bony operculum. The great majority of bony fish species have five pairs of gills, although a few have lost some over the course of evolution. The operculum can be important in adjusting the pressure of water inside of the pharynx to allow proper ventilation of the gills, so bony fish do not have to rely on ram ventilation (and hence near constant motion) to breathe. Valves inside the mouth keep the water from escaping.

The gill arches of bony fish typically have no septum, so the gills alone project from the arch, supported by individual gill rays. Some species retain gill rakers. Though all but the most primitive bony fish lack spiracles, the pseudobranch associated with them often remains, being located at the base of the operculum. This is, however, often greatly reduced, consisting of a small mass of cells without any remaining gill-like structure.

Marine teleosts also use gills to excrete electrolytes. The gills' large surface area tends to create a problem for fish that seek to regulate the osmolarity of their internal fluids. Salt water is less dilute than these internal fluids, so saltwater fish lose large quantities of water osmotically through their gills. To regain the water, they drink large amounts of sea water and excrete the salt. Fresh water is more dilute than the internal fluids of fish, however, so freshwater fish gain water osmotically through their gills.

Other Vertebrates

An Alpine newt larva showing the external gills, which flare just behind the head

Lampreys and hagfish do not have gill slits as such. Instead, the gills are contained in spherical pouches, with a circular opening to the outside. Like the gill slits of higher fish, each pouch contains two gills. In some cases, the openings may be fused together, effectively forming an operculum. Lampreys have seven pairs of pouches, while hagfishes may have six to fourteen, depending on the species. In the hagfish, the pouches connect with the pharynx internally and a separate tube which has no respiratory tissue (the pharyngocutaneous duct) develops beneath the pharynx proper, expelling ingested debris by closing a valve at its anterior end.

Tadpoles of amphibians have from three to five gill slits that do not contain actual gills. Usually no spiracle or true operculum is present, though many species have operculum-like structures. Instead of internal gills, they develop three feathery external gills that grow from the outer surface of the gill arches. Sometimes, adults retain these, but they usually disappear at metamorphosis. Lungfish larvae also have external gills, as does the primitive ray-finned fish *Polypterus*, though the latter has a structure different from amphibians. Some salamanders, such as the olm, and the mudpuppy, retain their external gills upon reaching adulthood.

Still, some extinct tetrapod groups did retain true gills. A study on *Archegosaurus* demonstrates that it had internal gills like true fish.

Branchia

Branchia (pl. branchiae) is the naturalists' name for gills. Galen observed that fish had multitudes of openings (*foramina*), big enough to admit gases, but too fine to give passage to water. Pliny the Elder held that fish respired by their gills, but observed that Aristotle was of another opinion. The word *branchia* comes from the Greek, gills, plural of (in singular, meaning a fin).

Invertebrate Gills

A live sea slug, *Pleurobranchaea meckelii*: The gill (or ctenidium) is visible in this view of the right-hand side of the animal.

Respiration in the echinoderms (such as starfish and sea urchins) is carried out using a very primitive version of gills called papulae. These thin protuberances on the surface of the body contain diverticula of the water vascular system. Crustaceans, molluscs, and some aquatic insects have tufted gills or plate-like structures on the surfaces of their bodies.

Caribbean hermit crabs have modified gills that allow them to live in humid conditions.

The gills of aquatic insects are tracheal, but the air tubes are sealed, commonly connected to thin external plates or tufted structures that allow diffusion. The oxygen in these tubes is renewed through the gills. In the larval dragon fly, the wall of the caudal end of the alimentary tract (rectum) is richly supplied with tracheae as a rectal gill, and water pumped into and out of the rectum provides oxygen to the closed tracheae.

Plastron

A plastron is a type of structural adaptation occurring among some aquatic arthropods (primarily insects), a form of inorganic gill which holds a thin film of atmospheric oxygen in an area with small openings called spiracles that connect to the tracheal system. The plastron typically consists of dense patches of hydrophobic setae on the body, which prevent water entry into the spiracles, but may also involve scales or microscopic ridges projecting from the cuticle. The physical properties of the interface between the trapped air film and surrounding water allow gas exchange through the spiracles, almost as if the insect were in atmospheric air. Carbon dioxide diffuses into the surrounding water due to its high solubility, while oxygen diffuses into the film as the concentration within the film has been reduced by respiration, and nitrogen also diffuses out as its tension has been increased. Oxygen diffuses into the air film at a higher rate than nitrogen diffuses out. However, water surrounding the insect can become oxygen-depleted if there is no water movement, so many such insects in still water actively direct a flow of water over their bodies.

The inorganic gill mechanism allows aquatic insects with plastrons to remain constantly submerged. Examples include many beetles in the family Elmidae, aquatic weevils, and true bugs in the family Aphelocheiridae, as well as at least one species of ricinuleid arachnid. A somewhat similar mechanism is used by the diving bell spider, which maintains an underwater bubble that exchanges gas like a plastron. Other diving insects (such as backswimmers, and hydrophilid beetles) may carry trapped air bubbles, but deplete the oxygen more quickly, and thus need constant replenishment.

Vision in Fishes

An oscar, *Astronotus ocellatus*, surveys its world

Vision is an important sensory system for most species of fish. Fish eyes are similar to the eyes of terrestrial vertebrates like birds and mammals, but have a more spherical lens. Birds and mammals (including humans) normally adjust focus by changing the shape of their lens, but fish normally adjust focus by moving the lens closer to or further from the retina. Fish retinas generally have both rod cells and cone cells (for scotopic and photopic vision), and most species have colour vision. Some fish can see ultraviolet and some are sensitive to polarized light.

Among jawless fish, the lamprey has well-developed eyes, while the hagfish has only primitive eyespots. The ancestors of modern hagfish, thought to be the protovertebrate were evidently pushed to very deep, dark waters, where they were less vulnerable to sighted predators, and where it is advantageous to have a convex eye-spot, which gathers more light than a flat or concave one. Fish vision shows evolutionary adaptation to their visual environment, for example deep sea fish have eyes suited to the dark environment.

Water as a Visual Environment

Fish and other aquatic animals live in a different light environment than terrestrial species. Water absorbs light so that with increasing depth the amount of light available decreases quickly. The optical properties of water also lead to different wavelengths of light being absorbed to different degrees. For example, visible light of long wavelengths (e.g. red, orange) is absorbed quicker than light of shorter wavelengths (green, blue). Ultraviolet light (even shorter wavelength than violet) is absorbed quicker yet. Besides these universal qualities of water, different bodies of water may absorb light of different wavelengths due to varying salt and/or chemical presence in the water.

Structure and Function

Fish eyes are broadly similar to those of other vertebrates – notably the tetrapods (amphibians, reptiles, birds and mammals – all of which evolved from a fish ancestor). Light enters the eye at the cornea, passing through the pupil to reach the lens. Most fish species seem to have a fixed pupil size, but elasmobranches (like sharks and rays) have a muscular iris which allows pupil diameter to be adjusted. Pupil shape varies, and may be e.g. circular or slit-like.

Lenses are normally spherical but can be slightly elliptical in some species. Compared to terrestrial vertebrates, fish lenses are generally more dense and spherical. In the aquatic environment there is not a major difference in the refractive index of the cornea and the surrounding water (compared to air on land) so the lens has to do the majority of the refraction. Due to "a refractive index gradient within the lens — exactly as one would expect from optical theory" the spherical lenses of fish are able to form sharp images free from spherical aberration.

Once light passes through the lens it is transmitted through a transparent liquid medium until it reaches the retina, containing the photoreceptors. Like other vertebrates, the photoreceptors are on the inside layer so light must pass through layers of other neurons before it reaches them. The retina contains rod cells and cone cells.

The Retina

Within the retina, rod cells provide high visual sensitivity (at the cost of acuity), being used in low light conditions. Cone cells provide higher spatial and temporal resolution than rods can, and allow for the possibility of colour vision by comparing absorbances across different types of cones which are more sensitive to different wavelengths. The ratio of rods to cones depends on the ecology of the fish species concerned, *e.g.*, those mainly active during the day in clear waters will have more cones than those living in low light environments. Colour vision is more useful in environments with a broader range of wavelengths available, *e.g.*, near the surface in clear waters rather than in deeper water where only a narrow band of wavelengths persist.

The distribution of photoreceptors across the retina is not uniform. Some areas have higher densities of cone cells, for example. Fish may have two or three areas specialized for high acuity (e.g. for prey capture) or sensitivity (e.g. from dim light coming from below). The distribution of photoreceptors may also change over time during development of the individual. This is especially the case when the species typically moves between different light environments during its life cycle (e.g. shallow to deep waters, or fresh water to ocean).

Some species have a tapetum, a reflective layer which bounces light that passes through the retina back through it again. This enhances sensitivity in low light conditions, such as nocturnal and deep sea species, by giving photons a second chance to be captured by photoreceptors. However this comes at a cost of reduced resolution. Some species are able to effectively turn their tapetum off in bright conditions, with a dark pigment layer covering it as needed.

The retina uses a lot of oxygen compared to most other tissues, and is supplied with plentiful oxygenated blood to ensure optimal performance.

Humans have a vestibulo-ocular reflex, which is a reflex eye movement that stabilizes images on the retina during head movement by producing an eye movement in the direction opposite to head movement, thus preserving the image on the center of the visual field. In a similar manner, fish have a vestibulo-ocular reflex which stabilizes visual images on the retina when it moves its tail.

Accommodation

Accommodation is the process by which the vertebrate eye adjusts focus on an object as it moves closer or further away. Whereas birds and mammals achieve accommodation by deforming the lens of their eyes, fish and amphibians normally adjust focus by moving the lens closer or further from the retina. They use a special muscle which changes the distance of the lens from the retina. In bony fishes the muscle is called the *retractor lentis*, and is relaxed for near vision, whereas for cartilaginous fishes the muscle is called the *protractor lentis*, and is relaxed for far vision. Thus bony fishes accommodate for distance vision by moving the lens further from the retina, while cartilaginous fishes accommodate for near vision by moving the lens closer to the retina.

Stabilising Images

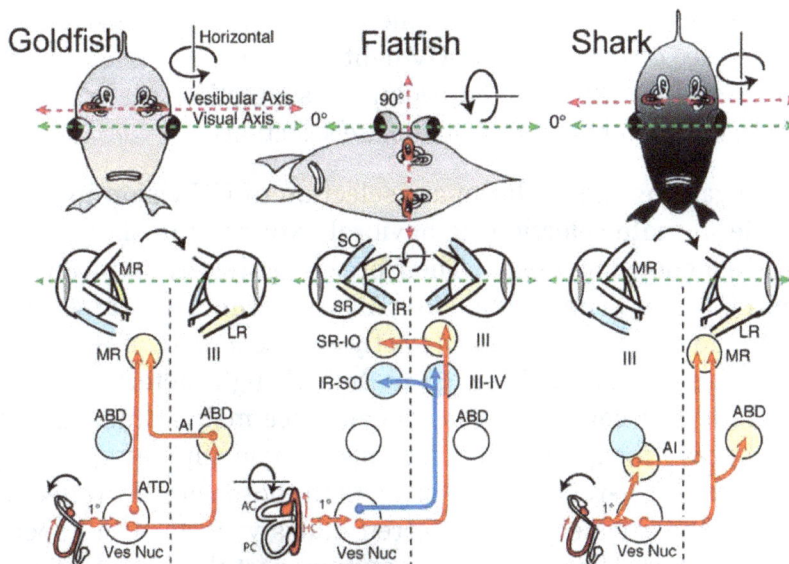

Horizontal vestibulo-ocular reflex in goldfish, flatfish and sharks

There is a need for some mechanism that stabilises images during rapid head movements. This is achieved by the vestibulo-ocular reflex, which is a reflex eye movement that stabilises images on the retina by producing eye movements in the direction opposite to head movements, thus preserving the image on the centre of the visual field. For example, when the head moves to the right, the eyes move to the left, and vice versa. In many animals, including human beings, the inner ear functions as the biological analogue of an accelerometer in camera image stabilization systems, to stabilize the image by moving the eyes. When a rotation of the head is detected, an inhibitory signal is sent to the extraocular muscles on one side and an excitatory signal to the muscles on the other side. The result is a compensatory movement of the eyes. Typical human eye movements lag head movements by less than 10 ms.

The diagram on the right shows the horizontal vestibulo-ocular reflex circuitry in bony and cartilaginous fish.

- "Goldfish" shows the principal three-neuronal vestibulo-ocular reflex linking the horizontal semicircular canal with contralateral abducens (ABD) and ipsilateral MR motoneurons.

- "Flatfish" shows that after 90° displacement of the vestibular relative to visual axis (metamorphosis) compensatory eye movements are produced by redirecting horizontal canal signals to vertical and oblique motoneurons.

- In "Shark" horizontal canal/second order neurons project to contralateral ABD and MR motoneurons including ipsilateral AI neurons. 1°, first order vestibular neuron; ATD, Ascending Tract of Deiter's.

Ultraviolet

Fish vision is mediated by four visual pigments that absorb various wavelengths of light. Each pigment is constructed from a chromophore and the transmembrane protein, known as opsin. Mutations in opsin have allowed for visual diversity, including variation in wavelength absorption. A mutation of the opsin on the SWS-1 pigment allows some vertebrates to absorb UV light ($\approx$360 nm), so they can see objects to reflect UV light. A wide range of fish species has developed and maintained this visual trait throughout evolution, suggesting it is advantageous. UV vision may be related to foraging, communication, and mate selection.

The leading theory regarding the evolutionary selection of UV vision in select fish species is due to its strong role in mate selection. Behavioral experiments show that African cichlids utilize visual cues when choosing a mate. Their breeding sites are typically in shallow waters with high clarity and UV light penetration. Male African cichlids are largely a blue color that is reflective in UV light. Females are able to correctly choose a mate of their species when these reflective visual cues are present. This suggests that UV light detection is crucial for correct mate selection. UV reflective color patterns also enhance male attractiveness in guppies and three-spined sticklebacks. In experimental settings, female guppies spent significantly more time inspecting males with UV-reflective coloring than those with UV reflection blocked. Similarly, female three-spined sticklebacks preferred males viewed in full spectrum over those viewed in UV blocking filters. These results strongly suggest the role of UV detection in sexual selection and, thus, reproductive fitness. The prominent role of UV light detection in fish mate choice has allowed the trait to be maintained over time. UV vision may also be related to foraging and other communication behaviors.

Many species of fish can see the ultraviolet end of the spectrum, beyond the violet.

Ultraviolet vision is sometimes used during only part of the life cycle of a fish. For example, juvenile brown trout live in shallow water where they use ultraviolet vision to enhance their ability to detect zooplankton. As they get older, they move to deeper waters where there is little ultraviolet light.

The two stripe damselfish, *Dascyllus reticulatus*, has ultraviolet-reflecting colouration which they appear to use as an alarm signal to other fish of their species. Predatory species cannot see this if their vision is not sensitive to ultraviolet. There is further evidence for this view that some fish use ultraviolet as a "high-fidelity secret communication channel hidden from predators", while yet other species use ultraviolet to make social or sexual signals.

Polarized Light

It is not easy to establish whether a fish is sensitive to polarized light, though it appears likely in a number of taxa. It has been unambiguously demonstrated in anchovies. The ability to detect polarized light may provide better contrast and/or directional information for migrating species. Polarized light is most abundant at dawn and dusk. Polarized light reflected from the scales of a fish may enable other fish to better detect it against a diffuse background, and may provide useful information to schooling fish about their proximity and orientation relative to neighbouring fish.

Double Cones

Most fish have double cones, a pair of cone cells joined to each other. Each member of the double cone may have a different peak absorbance, and behavioural evidence supports the idea that each type of individual cone in a double cone can provide separate information (i.e. the signal from individual members of the double cone are not necessarily summed together).

Adaptation to Habitat

Four-eyed fish
The four-eyed fish feeds at the surface of the water with eyes that allow it to see both above and below the surface at the same time.

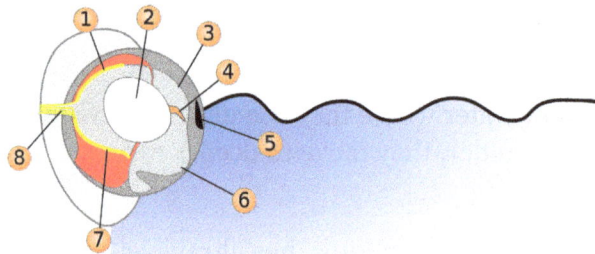

Eye of a four-eyed fish
1) Underwater retina 2) Lens 3) Air pupil 4) Tissue band 5) Iris 6) Underwater pupil 7) Air retina 8) Optic nerve

Fishes that live in surface waters down to about 200 metres, epipelagic fishes, live in a sunlit zone where visual predators use visual systems which are designed pretty much as might be expected. But even so, there can be unusual adaptations. Four-eyed fish have eyes raised above the top of the head and divided in two different parts, so that they can see below and above the water surface at the same time. Four-eyed fish actually have only two eyes, but their eyes are specially adapted for their surface-dwelling lifestyle. The eyes are positioned on the top of the head, and the fish floats at the water surface with only the lower half of each eye underwater. The two halves are divided by a band of tissue and the eye has two pupils, connected by part of the iris. The upper half of the eye is adapted for vision in air, the lower half for vision in water. The lens of the eye changes in thickness top to bottom to account for the difference in the refractive indices of air versus water. These fish spend most of their time at the surface of the water. Their diet mostly consists of the terrestrial insects which are available at the surface.

Deepwater fishes, like this Antarctic toothfish, often have large, upward looking eyes, adapted to detect prey silhouetted against the gloom above.

The telescopefish has large, forward-pointing telescoping eyes with large lenses.

The mesopelagic sabertooth is an ambush predator with telescopic, upward-pointing eyes.

Mesopelagic fishes live in deeper waters, in the twilight zone down to depths of 1000 metres, where the amount of sunlight available is not sufficient to support photosynthesis. These fish are adapted for an active life under low light conditions. Most of them are visual predators with large eyes. Some of the deeper water fish have tubular eyes with big lenses and only rod cells that look upwards. These give binocular vision and great sensitivity to small light signals. This adaptation gives improved terminal vision at the expense of lateral vision, and allows the predator to pick out squid, cuttlefish, and smaller fish that are silhouetted against the gloom above them. For more sensitive vision in low light, some fish have a retroreflector behind the retina. Flashlight fish have this plus photophores, which they use in combination to detect eyeshine in other fish.

Still deeper down the water column, below 1000 metres, are found the bathypelagic fishes. At this depth the ocean is pitch black, and the fish are sedentary, adapted to outputting minimum energy in a habitat with very little food and no sunlight. Bioluminescence is the only light available at these depths. This lack of light means the organisms have to rely on senses other than vision. Their eyes are small and may not function at all.

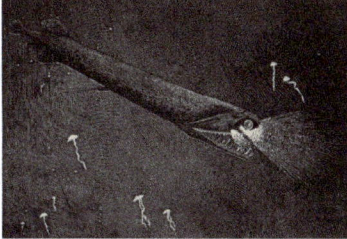

Most deep-sea fish cannot see red light. The deepwater stoplight loosejaw produces red bioluminescence so it can hunt with an effectively invisible beam of light.

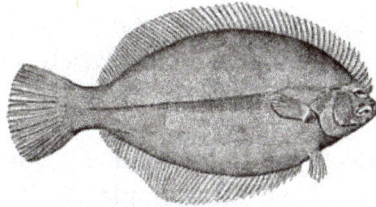

When the larvae of a flatfish grows, the eye on one side rotates to the other side so the fish can rest on the seafloor

The European plaice is a flatfish with raised eyes, so when it buries itself in sand for camouflage it can still see

At the very bottom of the ocean flatfish can be found. Flatfish are benthic fish with a negative buoyancy so they can rest on the seafloor. Although flatfish are bottom dwellers, they are not usually deep sea fish, but are found mainly in estuaries and on the continental shelf. When flatfish larvae hatch they have the elongated and symmetric shape of a typical bony fish. The larvae do not dwell on the bottom, but float in the sea as plankton. Eventually they start metamorphosing into the adult form. One of the eyes migrates across the top of the head and onto the other side of the body, leaving the fish blind on one side. The larva loses its swim bladder and spines, and sinks to the bottom, laying its blind side on the underlying surface. Richard Dawkins explains this as an example of evolutionary adaptation

...bony fish as a rule have a marked tendency to be flattened in a vertical direction.... It was natural, therefore, that when the ancestors of [flatfish] took to the sea bottom, they should have lain on one *side*.... But this raised the problem that one eye was always looking down into the sand and was effectively useless. In evolution this problem was solved by the lower eye 'moving' round to the upper side.

Prey usually have eyes on the sides of their head so they have a large field of view, from which to avoid predators. Predators usually have eyes in front of their head so they have better depth perception. Benthic predators, like flatfish, have eyes arranged so they have a binocular view of what is above them as they lie on the bottom.

Coloration

Fish have evolved sophisticated ways of using colouration. For example, prey fish have ways of using colouration to make it more difficult for visual predators to see them. In pelagic fish, these adaptations are mainly concerned with a reduction in silhouette, a form of camouflage. One method of achieving this is to reduce the area of their shadow by lateral compression of the body. Another method, also a form of camouflage, is by countershading in the case of epipelagic fish and by counter-illumination in the case of mesopelagic fish. Countershading is achieved by colouring the fish with darker pigments at the top and lighter pigments at the bottom in such a way that the colouring matches the background. When seen from the top, the darker dorsal area of the animal blends into the darkness of the water below, and when seen from below, the lighter ventral area blends into the sunlight from the surface. Counter illumination is achieved via bioluminescence by the production of light from ventral photophores, aimed at matching the light intensity from the underside of the fish with the light intensity from the background.

Benthic fish, which rest on the seafloor, physically hide themselves by burrowing into sand or retreating into nooks and crannies, or camouflage themselves by blending into the background or by looking like a rock or piece of seaweed.

While these tools may be effective as predator avoidance mechanisms, they also serve as equally effective tools for the predators themselves. For example, the deepwater velvet belly lantern shark uses counter-illumination to hide from its prey.

The foureye butterflyfish has false eyes on its back end, confusing predators about which is the front end of the fish

John Dory

Some fish species also display false eyespots. The foureye butterflyfish gets its name from a large dark spot on the rear portion of each side of the body. This spot is surrounded by a brilliant white ring, resembling an eyespot. A black vertical bar on the head runs through the true eye, making it hard to see. This can result in a predator thinking the fish is bigger than it is, and confusing the back end with the front end. The butterflyfish's first instinct when threatened is to flee, putting the false eyespot closer to the predator than the head. Most predators aim for the eyes, and this false eyespot tricks the predator into believing that the fish will flee tail first.

The John Dory is a benthopelagic coastal fish with a high laterally compressed body. Its body is so thin that it can hardly be seen from the front. It also has a large dark spot on both sides, which is used to flash an "evil eye" if danger approaches. The large eyes at the front of the head provide it with the bifocal vision and depth perception it needs to catch prey. The John Dory's eye spot on the side of its body also confuses prey, which is then scooped up in its mouth.

Barreleyes

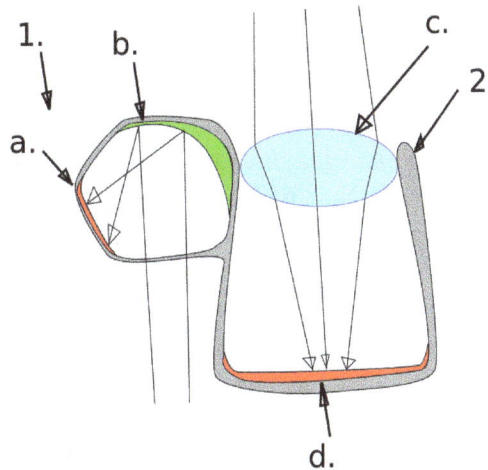

Left: The barreleye has barrel-shaped, telescopic eyes which are
generally directed upwards, but can also be swivelled forward

----

Right: The brownsnout spookfish is the only vertebrate known to employ a mirror eye (as well as a lens):
(1) diverticulum (2) main eye (a) retina (b) reflective crystals (c) lens (d) retina

Barreleyes are a family of small, unusual-looking mesopelagic fishes, named for their barrel-shaped, tubular eyes which are generally directed upwards to detect the silhouettes of available prey. Barreleyes have large, telescoping eyes which dominate and protrude from the skull. These eyes generally gaze upwards, but can also be swivelled forwards in some species. Their eyes have a large lens and a retina with an exceptional number of rod cells and a high density of rhodopsin (the "visual purple" pigment); there are no cone cells.

The barreleye species, *Macropinna microstoma*, has a transparent protective dome over the top of its head, somewhat like the dome over an airplane cockpit, through which the lenses of its eyes can be seen. The dome is tough and flexible, and presumably protects the eyes from the nematocysts (stinging cells) of the siphonophores from which it is believed the barreleye steals food.

Another barreleye species, the brownsnout spookfish, is the only vertebrate known to employ a mirror, as opposed to a lens, to focus an image in its eyes. It is unusual in that it utilizes both refractive and reflective optics to see. The main tubular eye contains a lateral ovoid swelling called a diverticulum, largely separated from the eye by a septum. The retina lines most of the interior of the eye, and there are two corneal openings, one directed up and the other down, that allow light into the main eye and the diverticulum respectively. The main eye employs a lens to focus its image, as in other fishes. However, inside the diverticulum the light is reflected and focused onto the retina by a curved composite mirror derived from the retinal tapetum, composed of many layers of small reflective plates possibly made of guanine crystals. The split structure of the brownsnout spookfish eye allows the fish to see both up and down at the same time. In addition, the mirror system is superior to a lens in gathering light. It is likely that the main eye serves to detect objects silhouetted against the sunlight, while the diverticulum serves to detect bioluminescent flashes from the sides and below.

Sharks

Shark eyes are similar to the eyes of other vertebrates, including similar lenses, corneas and retinas, though their eyesight is well adapted to the marine environment with the help of a tissue called tapetum lucidum. This tissue is behind the retina and reflects light back to it, thereby increasing visibility in the dark waters. The effectiveness of the tissue varies, with some sharks having stronger nocturnal adaptations. Many sharks can contract and dilate their pupils, like humans, something no teleost fish can do. Sharks have eyelids, but they do not blink because the surrounding water cleans their eyes. To protect their eyes some species have nictitating membranes. This membrane covers the eyes while hunting and when the shark is being attacked. However, some species, including the great white shark (*Carcharodon carcharias*), do not have this membrane, but instead roll their eyes backwards to protect them when striking prey. The importance of sight in shark hunting behavior is debated. Some believe that electro- and chemoreception are more significant, while others point to the nictating membrane as evidence that sight is important. Presumably, the shark would not protect its eyes were they unimportant. The use of sight probably varies with species and water conditions. The shark's field of vision can swap between monocular and stereoscopic at any time. A micro-spectrophotometry study of 17 species of shark found 10 had only rod photoreceptors and no cone cells in their retinas giving them good night vision while making them colorblind. The remaining seven species had in addition to rods a single type of cone photoreceptor sensitive to green and, seeing only in shades of grey and green, are believed to be effectively colorblind. The study indicates that an object's contrast against the background, rather than colour, may be more important for object detection.

Other Examples

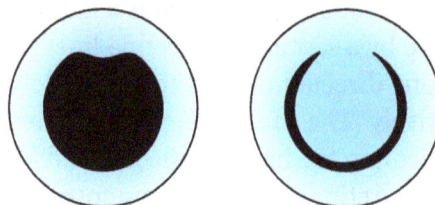

The omega iris allows Loricariids to adjust the amount of light that enters their eye

Small fish often school together for safety. This can have visual advantages, both by visually confusing predator fishes, and by providing many eyes for the school regarded as a body. The "predator confusion effect" is based on the idea that it becomes difficult for predators to pick out individual prey from groups because the many moving targets create a sensory overload of the predator's visual channel. "Shoaling fish are the same size and silvery, so it is difficult for a visually oriented predator to pick an individual out of a mass of twisting, flashing fish and then have enough time to grab its prey before it disappears into the shoal." The "many eyes effect" is based on the idea that as the size of the group increases, the task of scanning the environment for predators can be spread out over many individuals, a mass collaboration presumably providing a higher level of vigilance.

Fish are normally cold-blooded, with body temperatures the same as the surrounding water. However, some oceanic predatory fish, such as swordfish and some shark and tuna species, can warm parts of their body when they hunt for prey in deep and cold water. The highly visual swordfish uses a heating system involving its muscles which raises the temperature in its eyes and brain by up to 15 °C. The warming of the retina improves the rate at which the eyes respond to changes in rapid motion made by its prey by as much as ten times.

Some fish have eyeshine. Eyeshine is the result of a light-gathering layer in the eyes called the tapetum lucidum, which reflects white light. It does not occur in humans, but can be seen in other species, such as deer in a headlight. Eyeshine allows fish to see well in low-light conditions as well as in turbid (stained or rough, breaking) waters, giving them an advantage over their prey. This enhanced vision allows fish to populate the deeper regions in the ocean or a lake. In particular, freshwater walleye are so named because their eyeshine.

Many species of Loricariidae, a family of catfish, have a modified iris called an *omega iris*. The top part of the iris descends to form a loop which can expand and contract called an iris operculum; when light levels are high, the pupil reduces in diameter and the loop expands to cover the center of the pupil giving rise to a crescent shaped light transmitting portion. This feature gets its name from its similarity to an upside-down Greek letter omega (Ω). The origins of this structure are unknown, but it has been suggested that breaking up the outline of the highly visible eye aids camouflage in what are often highly mottled animals.

Distance Sensory Systems

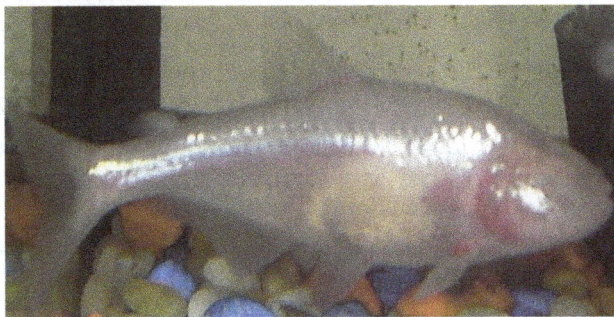

Blind cavefish find their way around by means of lateral lines, which are highly sensitive to changes in water pressure.

Visual systems are distance sensory systems which provide fish with data about location or objects at a distance without a need for the fish to directly touch them. Such distance sensing systems are important, because they allow communication with other fish, and provide information about the location of food and predators, and about avoiding obstacles or maintaining position in fish schools. For example, some schooling species have "schooling marks" on their sides, such as visually prominent stripes which provide reference marks and help adjacent fish judge their relative positions. But the visual system is not the only one that can perform such functions. Some schooling fish also have a lateral line running the length of their bodies. This lateral line enables the fish to sense changes in water pressure and turbulence adjacent to its body. Using this information, schooling fish can adjust their distance from adjacent fish if they come too close or stray too far.

The visual system in fish is augmented by other sensing systems with comparable or complimentary functions. Some fish are blind, and must rely entirely on alternate sensing systems. Other senses which can also provide data about location or distant objects include hearing and echolocation, electroreception, magnetoception and chemoreception (smell and taste). For example, catfish have chemoreceptors across their entire bodies, which means they "taste" anything they touch and "smell" any chemicals in the water. "In catfish, gustation plays a primary role in the orientation and location of food".

Cartilaginous fish (sharks, stingrays and chimaeras) use magnetoception. They possess special electroreceptors called the *ampullae of Lorenzini* which detect a slight variation in electric potential. These receptors, located along the mouth and nose of the fish, operate according to the principle that a time-varying magnetic field moving through a conductor induces an electric potential across the ends of the conductor. The ampullae may also allow the fish to detect changes in water temperature. As in birds, magnetoception may provide information which help the fish map migration routes.

References

- Kris S. Freeman (January 2012). "Remediating Soil Lead with Fishbones". Environmental Health Perspectives. 120 (1): a20–a21. PMC 3261960. PMID 22214821. doi:10.1289/ehp.120-a20a

- Kuraku; Hoshiyama, D; Katoh, K; Suga, H; Miyata, T; et al. (December 1999). "Monophyly of Lampreys and Hagfishes Supported by Nuclear DNA–Coded Genes". Journal of Molecular Evolution. 49 (6): 729–35. PMID 10594174. doi:10.1007/PL00006595

- Romer, Alfred Sherwood; Parsons, Thomas S. (1977). The Vertebrate Body. Philadelphia, PA: Holt-Saunders International. pp. 173–177. ISBN 0-03-910284-X

- Muller, M. (1996). "A novel classification of planar four-bar linkages and its application to the mechanical analysis of animal systems" (PDF). Phil. Trans. R. Soc. Lond. B. 351 (1340): 689–720. doi:10.1098/rstb.1996.0065

- Farrell, Anthony P, ed. (1 June 2011). Encyclopedia of Fish Physiology: From Genome to Environment. Stevens, E Don; Cech, Jr., Joseph J; Richards, Jeffrey G. Academic Press. p. 2315. ISBN 978-0-08-092323-9

- Stock, David; Whitt GS (7 August 1992). "Evidence from 18S ribosomal RNA sequences that lampreys and hagfishes form a natural group". Science. 257 (5071): 787–9. Bibcode:1992Sci...257..787S. PMID 1496398. doi:10.1126/science.1496398. Retrieved 22 November 2011

- Bell CC, Han V, Sawtell NB (2008). "Cerebellum-like structures and their implications for cerebellar function". Annu. Rev. Neurosci. 31: 1–24. PMID 18275284. doi:10.1146/annurev.neuro.30.051606.094225

- Romer, Alfred Sherwood; Parsons, Thomas S. (1977). The Vertebrate Body. Philadelphia, PA: Holt-Saunders International. pp. 385–386. ISBN 0-03-910284-X

- Briggs, John C. (2005). "The biogeography of otophysian fishes (Ostariophysi: Otophysi): a new appraisal" (PDF). Journal of Biogeography. 32 (2): 287–294. doi:10.1111/j.1365-2699.2004.01170.x

- Woodhams PL (1977). "The ultrastructure of a cerebellar analogue in octopus". J Comp Neurol. 174 (2): 329–45. PMID 864041. doi:10.1002/cne.901740209

- Romer, Alfred Sherwood; Parsons, Thomas S. (1977). The Vertebrate Body. Philadelphia, PA: Holt-Saunders International. pp. 316–327. ISBN 0-03-910284-X

- Shi Z, Zhang Y, Meek J, Qiao J, Han VZ (2008). "The neuronal organization of a unique cerebellar specialization: the valvula cerebelli of a mormyrid fish". J. Comp. Neurol. 509 (5): 449–73. PMID 18537139. doi:10.1002/cne.21735

4

Fish Locomotion: An Integrated Study

Fish locomotion is the movement of a fish. The different forms of locomotion are anguilliform, sub-carangiform, carangiform and rapid swimming. The aspects elucidated in this chapter are of vital importance, and provide a better understanding of fish locomotion.

Fish Locomotion

Fish propel themselves through water using many different mechanisms.

Fish locomotion is the variety of types of animal locomotion used by fish, principally by swimming. This however is achieved in different groups of fish by a variety of mechanisms of propulsion in water, most often by wavelike movements of the fish's body and tail, and in various specialised fish by movements of the fins. The major forms of locomotion in fish are anguilliform, in which a wave passes evenly along a long slender body; sub-carangiform, in which the wave increases quickly in amplitude towards the tail; carangiform, in which the wave is concentrated near the tail, which oscillates rapidly; thunniform, rapid swimming with a large powerful crescent-shaped tail; and ostraciiform, with almost no oscillation except of the tail fin. More specialised fish include movement by pectoral fins with a mainly stiff body, as in the sunfish; and movement by propagating a wave along the long fins with a motionless body in fish with electric organs such as the knifefish.

In addition, some fish can variously "walk", i.e., move over land, burrow in mud, and glide through the air.

Swimming

Fish swim by exerting force against the surrounding water. There are exceptions, but this is normally achieved by the fish contracting muscles on either side of its body in order to generate waves of flexion that travel the length of the body from nose to tail, generally getting larger as they go along. The vector forces exerted on the water by such motion cancel out laterally, but generate a net force backwards which in turn pushes the fish forward through the water. Most fishes generate thrust using lateral movements of their body and caudal fin, but many other species move mainly using their median and paired fins. The latter group swim slowly, but can turn rapidly, as is needed when living in coral reefs for example. But they can't swim as fast as fish using their bodies and caudal fins.

Body/Caudal Fin Propulsion

There are five groups that differ in the fraction of their body that is displaced laterally:

Anguilliform

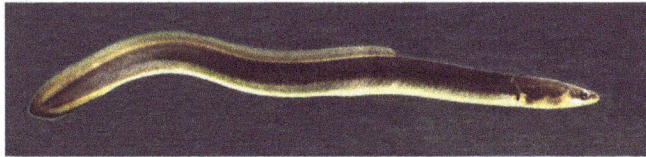

Eels propagate a more or less constant-sized flexion wave along their slender bodies.

In the anguilliform group, containing some long, slender fish such as eels, there is little increase in the amplitude of the flexion wave as it passes along the body.

Sub-carangiform

The sub-carangiform group has a more marked increase in wave amplitude along the body with the vast majority of the work being done by the rear half of the fish. In general, the fish body is stiffer, making for higher speed but reduced maneuverability. Trout use sub-carangiform locomotion.

Carangiform

The carangiform group, named for the Carangidae, are stiffer and faster-moving than the previous groups. The vast majority of movement is concentrated in the very rear of the body and tail. Carangiform swimmers generally have rapidly oscillating tails.

Thunniform

The thunniform group contains high-speed long-distance swimmers, and is a unique trait (an autapomorphy) of the tunas. Here, virtually all the sideways movement is in the tail and the region connecting the main body to the tail (the peduncle). The tail itself tends to be large and crescent shaped.

Ostraciiform

The ostraciiform group have no appreciable body wave when they employ caudal locomotion. Only the tail fin itself oscillates (often very rapidly) to create thrust. This group includes Ostraciidae.

Median/Paired Fin Propulsion

Boxfish use median-paired fin swimming, as they are not well streamlined,
and use primarily their pectoral fins to produce thrust.

Not all fish fit comfortably in the above groups. Ocean sunfish, for example, have a completely different system, the tetraodontiform mode, and many small fish use their pectoral fins for swimming as well as for steering and dynamic lift. Fish with electric organs, such as those in the knifefish (Gymnotiformes), swim by undulating their very long fins while keeping the body still, presumably so as not to disturb the electric field that they generate.

Many fish swim using combined behavior of their two pectoral fins or both their anal and dorsal fins. Different types of Median paired fin propulsion can be achieved by preferentially using one fin pair over the other, and include rajiform, diodontiform, amiiform, gymnotiform and balistiform modes.

Rajiform

Rajiform locomotion is characteristic of rays, skates, and mantas when thrust is produced by vertical undulations along large, well developed pectoral fins.

Diodontiform

Porcupine fish (here, *Diodon nicthemerus*) swim by undulating their pectoral fins.

Diodontiform locomotion propels the fish propagating undulations along large pectoral fins, as seen in the porcupinefish (Diodontidae).

Amiiform

Amiiform locomotion consists of undulations of a long dorsal fin while the body axis is held straight and stable, as seen in the bowfin.

Gymnotiform

Gymnotus maintains a straight back while swimming to avoid disturbing its electric sense.

Gymnotiform locomotion consists of undulations of a long anal fin, essentially upside down amiiform, seen in the knifefish (Gymnotiformes).

Balistiform

In balistiform locomotion, both anal and dorsal fins undulate, as seen in the Zeidae.

Oscillatory

Oscillation is viewed as pectoral-fin-based swimming and is best known as mobuliform locomotion. The motion can be described as the production of less than half a wave on the fin, similar to a bird wing flapping. Pelagic stingrays, such as the manta, cownose, eagle and bat rays use oscillatory locomotion.

Tetraodontiform

In tetraodontiform locomotion, the dorsal and anal fins are flapped as a unit, either in phase or exactly opposing one another, as seen in the Tetraodontiformes (boxfishes and pufferfishes). The ocean sunfish displays an extreme example of this mode.

Labriform

In labriform locomotion, seen in the wrasses (Labriformes), oscillatory movements of pectoral fins are either drag based or lift based. Propulsion is generated either as a reaction to drag produced by dragging the fins through the water in a rowing motion, or via lift mechanisms.

Dynamic Lift

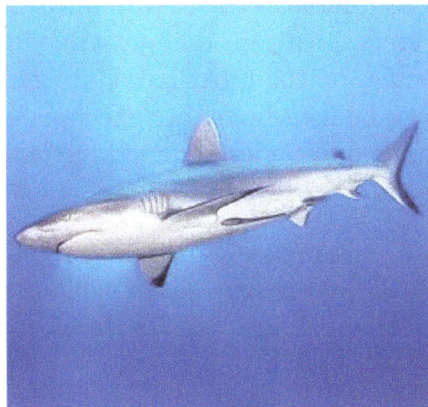

Sharks are denser than water, and must swim continually, using dynamic lift from their pectoral fins.

Bone and muscle tissues of fish are denser than water. To maintain depth fish such as sharks, but also some bony fish, increase buoyancy by means of a gas bladder or by storing oils or lipids. Fish without these features use dynamic lift instead. It is done using their pectoral fins in a manner similar to the use of wings by airplanes and birds. As these fish swim, their pectoral fins are positioned to create lift which allows the fish to maintain a certain depth. The two major drawbacks of this method are that these fish must stay moving to stay afloat and that they are incapable of swimming backwards or hovering.

Hydrodynamics

Similarly to the aerodynamics of flight, powered swimming requires animals to overcome drag by producing thrust. Unlike flying, however, swimming animals often do not need to supply much vertical force because the effect of buoyancy can counter the downward pull of gravity, allowing these animals to float without much effort. While there is great diversity in fish locomotion, swimming behavior can be classified into two distinct "modes" based on the body structures involved in thrust production, Median-Paired Fin (MPF) and Body-Caudal Fin (BCF). Within each of these classifications, there are numerous specifications along a spectrum of behaviours from purely undulatory to entirely oscillatory. In undulatory swimming modes, thrust is produced by wave-like movements of the propulsive structure (usually a fin or the whole body). Oscillatory modes, on the other hand, are characterized by thrust produced by swiveling of the propulsive structure on an attachment point without any wave-like motion.

Body-Caudal Fin

Most fish swim by generating undulatory waves that propagate down the body through the caudal fin. This form of undulatory locomotion is termed Body-Caudal Fin (BCF) swimming on the basis of the body structures used; it includes anguilliform, sub-carangiform, carangiform, and thunniform locomotory modes, as well as the oscillatory ostraciiform mode.

Adaptation

Similar to adaptation in avian flight, swimming behaviors in fish can be thought of as a balance of stability and maneuverability. Because BCF swimming relies on more caudal body structures that can direct powerful thrust only rearwards, this form of locomotion is particularly effective for accelerating quickly and cruising continuously. BCF swimming is, therefore, inherently stable and is often seen in fish with large migration patterns that must maximize efficiency over long periods. Propulsive forces in MPF swimming, on the other hand, are generated by multiple fins located on either side of the body that can be coordinated to execute elaborate turns. As a result, MPF swimming is well adapted for high maneuverability and is often seen in smaller fish that require elaborate escape patterns.

The habitats occupied by fishes are often related to their swimming capabilities. On coral reefs, the faster-swimming fish species typically live in wave-swept habitats subject to fast water flow speeds, while the slower fishes live in sheltered habitats with low levels of water movement.

Fish do not rely exclusively on one locomotor mode, but are rather locomotor generalists, choosing among and combining behaviors from many available behavioral techniques. At slower speeds, predominantly BCF swimmers often incorporate movement of their pectoral, anal, and dorsal fins as an additional stabilizing mechanism at slower speeds, but hold them close to their body at high speeds to improve streamlining and reducing drag. Zebrafish have even been observed to alter their locomotor behavior in response to changing hydrodynamic influences throughout growth and maturation.

In addition to adapting locomotor behavior, controlling buoyancy effects is critical for aquatic survival since aquatic ecosystems vary greatly by depth. Fish generally control their depth by regulating the amount of gas in specialized organs that are much like balloons. By changing the amount of gas in these swim bladders, fish actively control their density. If they increase the amount of air in their swim bladder, their overall density will become less than the surrounding water, and increased upward buoyancy pressures will cause the fish to rise until they reach a depth at which they are again at equilibrium with the surrounding water.

Flight

The transition of predominantly swimming locomotion directly to flight has evolved in a single family of marine fish, the Exocoetidae. Flying fish are not true fliers in the sense that they do not execute powered flight. Instead, these species glide directly over the surface of the ocean water without ever flapping their "wings." Flying fish have evolved abnormally large pectoral fins that act as airfoils and provide lift when the fish launches itself out of the water. Additional forward thrust and steering forces are created by dipping the hypocaudal (i.e. bottom) lobe of their caudal fin into the water and vibrating it very quickly, in contrast to diving birds in which these forces are produced by the same locomotor module used for propulsion. Of the 64 extant species of flying fish, only two distinct body plans exist, each of which optimizes two different behaviors.

Flying fish gain sufficient lift to glide above the water thanks to their enlarged pectoral fins.

Tradeoffs

While most fish have caudal fins with evenly sized lobes (i.e. homocaudal), flying fish have an enlarged ventral lobe (i.e. hypocaudal) which facilitates dipping only a portion of the tail back onto the water for additional thrust production and steering.

Because flying fish are primarily aquatic animals, their body density must be close to that of water for buoyancy stability. This primary requirement for swimming, however, means that flying fish are heavier (have a larger mass) than other habitual fliers, resulting in higher wing loading and lift to drag ratios for flying fish compared to a comparably sized bird. Differences in wing area, wing span, wing loading, and aspect ratio have been used to classify flying fish into two distinct classifications based on these different aerodynamic designs.

Biplane Body Plan

In the biplane or *Cypselurus* body plan, both the pectoral and pelvic fins are enlarged to provide lift during flight. These fish also tend to have "flatter" bodies which increase the total lift producing area thus allowing them to "hang" in the air better than more streamlined shapes. As a result of this high lift production, these fish are excellent gliders and are well adapted for maximizing flight distance and duration.

Comparatively, *Cypselurus* flying fish have lower wing loading and smaller aspect ratios (i.e. broader wings) than their *Exocoetus* monoplane counterparts, which contributes to their ability to fly for longer distances than fish with this alternative body plan. Flying fish with the biplane design take advantage of their high lift production abilities when launching from the water by utilizing a "taxiing glide" in which the hypocaudal lobe remains in the water to generate thrust even after the trunk clears the water's surface and the wings are opened with a small angle of attack for lift generation.

In the monoplane body plan of *Exocoetus*, only the pectoral fins are abnormally large, while the pelvic fins are small.

Monoplane Body Plan

In the *Exocoetus* or monoplane body plan, only the pectoral fins are enlarged to provide lift. Fish with this body plan tend to have a more streamlined body, higher aspect ratios (long, narrow wings), and higher wing loading than fish with the biplane body plan, making these fish well adapted for higher flying speeds. Flying fish with a monoplane body plan demonstrate different launching behaviors from their biplane counterparts. Instead of extending their duration of thrust production, monoplane fish launch from the water at high speeds at a large angle of attack (sometimes up to 45 degrees). In this way, monoplane fish are taking advantage of their adaptation for high flight speed, while fish with biplane designs exploit their lift production abilities during takeoff.

Walking

A "walking fish" is a fish that is able to travel over land for extended periods of time. Some other cases of nonstandard fish locomotion include fish "walking" along the sea floor, such as the handfish or frogfish.

Most commonly, walking fish are amphibious fish. Able to spend longer times out of water, these fish may use a number of means of locomotion, including springing, snake-like lateral undulation, and tripod-like walking. The mudskippers are probably the best land-adapted of contemporary fish and are able to spend days moving about out of water and can even climb mangroves, although to only modest heights. The Climbing gourami is often specifically referred to as a "walking fish", although it does not actually "walk", but rather moves in a jerky way by supporting itself on the extended edges of its gill plates and pushing itself by its fins and tail. Some reports indicate that it can also climb trees.

There are a number of fish that are less adept at actual walking, such as the walking catfish. Despite being known for "walking on land", this fish usually wriggles and may use its pectoral fins to aid in its movement. Walking Catfish have a respiratory system that allows them to live out of water for several days. Some are invasive species. A notorious case in the United States is the Northern snakehead. Polypterids have rudimentary lungs and can also move about on land, though rather clumsily. The Mangrove rivulus can survive for months out of water and can move to places like hollow logs.

Ogcocephalus parvus

There are some species of fish that can "walk" along the sea floor but not on land; one such animal is the flying gurnard (it does not actually fly, and should not be confused with flying fish). The batfishes of the Ogcocephalidae family (not to be confused with Batfish of Ephippidae) are also capable of walking along the sea floor. Bathypterois grallator, also known as a "tripodfish", stands on its three fins on the bottom of the ocean and hunts for food. The African lungfish (*P. annectens*) can use its fins to *"walk"* along the bottom of its tank in a manner similar to the way amphibians and land vertebrates use their limbs on land.

Burrowing

Many fishes, particularly eel-shaped fishes such as true eels, moray eels, and spiny eels, are capable of burrowing through sand or mud. Ophichthids, the snake eels, are capable of burrowing either forwards or backwards.

Fin and Flipper Locomotion

Fin and flipper locomotion occurs mostly in aquatic locomotion, and rarely in terrestrial locomotion. From the three common states of matter — gas, liquid and solid, these appendages are adapted for liquids, mostly fresh or saltwater and used in locomotion, steering and balancing of the body. Locomotion is important in order to escape predators, acquire food, find mates and bury for shelter, nest or food. Aquatic locomotion consists of swimming, whereas terrestrial locomotion encompasses walking, 'crutching', jumping, digging as well as covering. Some animals such as sea turtles and mudskippers use these two environments for different purposes, for example using the land for nesting, and the sea to hunt for food.

A species of mudskipper
(*Periophthalmus gracilis*)

Aquatic Locomotion with Fins and Flippers

Aquatic Locomotion of Fish

Fish live in Fresh or Saltwater habitats and some exceptions are capable of coming on land (Mudskippers). Most fish have a line of muscle blocks, called myomeres, along each side of the body. To swim, they alternately contract one side and relax the other side in a progression which goes from the head to the tail. In this way, an undulatory locomotion results, first bending the body one way in a wave which travels down the body, and then back the other way, with the contracting and relaxing muscles switching roles. They use their fins to propel themselves through the water in this swimming motion. Actinopterygians, the ray-finned fish show an evolutionary pattern of fine control ability to control the dorsal and ventral lobe of the caudal fin. Through developmental changes, intrinsic caudal muscles were added, which enable fish to exhibit such complex maneuvers such as control during acceleration, braking and backing. Studies have shown that the muscles in the caudal fin, have independent activity patterns from the myotomal musculature. These results show specific kinematic roles for different part of the fish's musculature. A curious example of fish adaption is the Ocean sunfish, also known as the *Mola mola*. These fish have undergone significant developmental changes reducing their spinal cord, giving them a disk like appearance, and investing in two very large fins for propulsion. This adaptation usually gives them the appearance that they are as long as they are tall. They are also amazing fish in that they hold the world record in weight gain from fry to adult (60 million times its weight).

Aquatic Locomotion of Marine Mammals

Swimming mammals, such as whales, dolphins,and sea lions, use their flippers to move forward through the water column. During swimming sea lions have a thrust phase, which lasts about 60% of the full cycle, and the recovery phase lasts the remaining 40%. A full cycle duration lasts about 0.5 to 1.0 seconds. Changing direction is a very rapid maneuver that is initiated by head movement towards the back of the animal that is followed by a spiral turn with the body. Due to their pectoral flippers being so closely located to their center of gravity, sea lions are capable of displaying astounding maneuverability in the pitch, roll, and yaw direction and are therefore not constrained, turning stochastically as they please. It is hypothesized that the increased level of maneuverability is caused by their complex habitat. Hunting occurs in difficult environments containing rocky inshore/kelp forest communities, with many niches for prey to hide, therefore requiring speed and maneuverability for capture. The complex skills of a sea lion are learned early on in ontogeny and most are perfected by the time the pups reach one year. Whales and dolphins are less maneuverable and more constrained in their movements. However, dolphins are capable of accelerating as fast as sea lions, but they are not capable of turning as quickly and as efficiently. For both whales and dolphins, their center of gravity does not line up with their pectoral flippers in a straight line, causing a much more rigid and stable swimming pattern.

Aquatic Locomotion of Marine Reptiles

Aquatic reptiles such as sea turtles predominantly use their pectoral flippers to propulse through the water and their pelvic flippers for maneuvering. During swimming they move their pectoral flippers in a clapping motion underneath their body and pull them back up into an airplane position, causing forward motion. During the swimming motion it is really important that they rotate their front flipper in order to decrease drag through the water column and increase their efficiency. Sea turtles exhibit a natural suite of behavior skills that help them direct themselves towards the ocean as well as identify the transition from sand to water after hatching. If rotated in the pitch, yaw or roll direction the hatchlings are capable of counteracting the forces acting upon them by correcting with either their pectoral or pelvic flippers and redirecting themselves towards the open ocean.

Terrestrial Locomotion

Terrestrial Locomotion of Fish

Mudskippers in The Gambia

Terrestrial locomotion poses new obstacles such as gravity and new media, including sand, mudd, twigs, logs, debris, grass and many more. Fins and flippers are aquatically adapted appendages and typically aren't very useful in such an environment. It could be hypothesized that fish would try to "swim" on land, but studies have shown that some fish evolved to cope with the terrestrial environment. Mudskippers, for example demonstrate a 'crutching' gait which enables them to 'walk' over muddy surfaces as well as dig burrows to hide in. Mudskippers are also able to jump up to 3 cm distances. This behavior is described as starting with a J-curvature of the body at about 2/3 of its body length (with its tail wrapped towards the head), followed by a straightening of their body which propulses them like a projectile through the air. This behavior enables them to cope with the new environment and opens their habitat to new food sources as well as new predators.

Terrestrial Locomotion of Marine Reptiles

Caretta caretta Jekyll Island, GA

Reptiles, such as sea turtles spend most of their lives in the ocean. However, their life cycle requires the females to come on shore and lay their nests on the beach. Consequently, the hatchlings emerge from the sand and have to run toward the water. Depending on their species, sea turtles are described to have either a symmetrical gait (diagonally opposite limbs are moving together) or an asymmetrical gait (Contra-lateral limbs move together). For example, loggerhead sea turtle hatchlings are commonly seen exhibiting symmetrical gait on sand, whereas, leatherback sea turtles employ the asymmetrical gait while on land. Notably, leatherbacks employ their front (pelvic) flippers more during forward terrestrial locomotion. Sea turtles can be seen nesting on subtropical and tropical beaches all around the world and exhibit such behavior such as arribada (Collective animal behavior). This is a phenomenon seen in Kemp's Ridley turtles which emerge all at once in one night only onto the beach to lay their nests.

Amphibious Fish

Amphibious fish are fish that are able to leave water for extended periods of time. About 11 distantly related genera of fish are considered amphibious. This suggests that many fish genera independently evolved amphibious traits, a process known as convergent evolution. These fish use a range of terrestrial locomotory modes, such as lateral undulation, tripod-like walking (using paired fins and tail), and jumping. Many of these locomotory modes incorporate multiple combinations of pectoral, pelvic and tail fin movement.

Many ancient fish had lung-like organs, and a few, such as the lungfish, still do. Some of these ancient "lunged" fish were the ancestors of tetrapods. However, in most recent fish species these organs evolved into the swim bladders, which help control buoyancy. Having no lung-like organs, modern amphibious fish and many fish in oxygen-poor water use other methods such as their gills or their skin to breathe air. Amphibious fish may also have eyes adapted to allow them to see clearly in air, despite the density differences between air and water.

List of Amphibious Fish

Lung Breathers

- Lungfish (*Dipnoi*): Six species, have limb like fins, and can breathe air. Some are obligate air breathers, meaning they will drown if not given access to breathe air. Some species will bury in the mud when the body of water they live in dries up, surviving up to two years until water returns.

- Various other "lunged" fish: now extinct, a few of this group were ancestors of the stem tetrapods that led to all tetrapods: Lissamphibia, sauropsids and mammals.

Gill or Skin Breathers

- Rockskippers: These blennies are found in Panama and elsewhere on the western coastline of the Americas. These fish come onto land to catch prey and escape aquatic predators. They often come out of water for up to 20 minutes. Leaping blennies (*Alticus arnoldorum*) are able to jump over land using their tails.

- Woolly sculpin (*Clinocottus analis*): Found in tide pools along the Pacific coast, these sculpins will leave water if the oxygen levels get low and can breathe air.

- Mudskippers (Oxudercinae): This subfamily of gobies is probably the most land adapted of fish. Mudskippers are found in mangrove swamps in Africa and the Indo-Pacific, they frequently come onto land and can survive in air for up to three and a half days. Mudskippers breathe through their skin and through the lining of the mouth (the mucosa) and throat (the pharynx). This requires the mudskipper to be wet, limiting mudskippers to humid habitats. This mode of breathing, similar to that employed by amphibians, is known as cutaneous breathing. They propel themselves over land on their sturdy forefins.

- Eels: Some eels, such as the European eel and the American eel, can live for an extended time out of water and can crawl on land if the soil is moist.

- Snakehead fish (Channidae): This family of fish are obligate air breathers, breathing air using their suprabranchial organ, which is a primitive labyrinth organ. The northern snakehead of Southeast Asia can "walk" on land by wriggling and using its pectoral fins, which allows it to move between the slow-moving, and often stagnant and temporary bodies of water in which it lives.

- Airbreathing catfish (Clariidae): Amphibious species of this family may venture onto land in wet weather, such as the eel catfish (*Channallabes apus*), which lives in swamps in Africa, and is known to hunt beetles on land.

- Labyrinth fish (Anabantoidei). This suborder of fish also use a labyrinth organ to breathe air. Some species from this group can move on land. Amphibious fish from this family are the climbing perches, African and Southeast Asian fish that are capable of moving from pool to pool over land by using their pectoral fins, caudal peduncle and gill covers as a means of locomotion. It is said that climbing gourami move at night in groups.

Undulatory Locomotion

Snakes primarily rely on undulatory locomotion to move through a wide range of environments

Undulatory locomotion is the type of motion characterized by wave-like movement patterns that act to propel an animal forward. Examples of this type of gait include crawling in snakes, or swimming in the lamprey. Although this is typically the type of gait utilized by limbless animals, some creatures with limbs, such as the salamander, choose to forgo use of their legs in certain environments and exhibit undulatory locomotion. This movement strategy is important to study in order to create novel robotic devices capable of traversing a variety of environments.

Environmental Interactions

In limbless locomotion, forward locomotion is generated by propagating flexural waves along the length of the animal's body. Forces generated between the animal and surrounding environment lead to a generation of alternating sideways forces that act to move the animal forward. These forces generate thrust and drag.

Hydrodynamics

Simulation predicts that thrust and drag are dominated by viscous forces at low Reynolds numbers and inertial forces at higher Reynolds numbers. When the animal swims in a fluid, two main forces are thought to play a role:

- Skin Friction: Generated due to the resistance of a fluid to shearing and is proportional to speed of the flow. This dominates undulatory swimming in spermatozoa and the nematode

- Form Force: Generated by the differences in pressure on the surface of the body and it varies with the square of flow speed.

At low Reynolds number (Re~10^0), skin friction accounts for nearly all of the thrust and drag. For those animals which undulate at intermediate Reynolds number (Re~10^1), such as the Ascidian larvae, both skin friction and form force account for the production of drag and thrust. At high Reynolds number (Re~10^2), both skin friction and form force act to generate drag, but only form force produces thrust.

Kinematics

In animals that move without use of limbs, the most common feature of the locomotion is a rostral to caudal wave that travels down their body. However, this pattern can change based on the particular undulating animal, the environment, and the metric in which the animal is optimizing (i.e. speed, energy, etc.). The most common mode of motion is simple undulations in which lateral bending is propagated from head to tail.

Snakes can exhibit 5 different modes of terrestrial locomotion: (1) lateral undulation, (2) side-winding, (3) concertina, (4) rectilinear, and (5) slide-pushing. Lateral undulation closely resembles the simple undulatory motion observed in many other animals such as in lizards, eels and fish, in which waves of lateral bending propagate down the snake's body.

The American eel typically moves in an aquatic environment, though it can also move on land for short periods of time. It is able to successfully move about in both environments by producing traveling waves of lateral undulations. However, differences between terrestrial and aquatic locomotor strategy suggest that the axial musculature is being activated differently. In terrestrial locomotion, all points along the body move on approximately the same path and, therefore, the lateral displacements along the length of the eel's body is approximately the same. However, in aquatic locomotion, different points along the body follow different paths with increasing lateral amplitude more posteriorly. In general, the amplitude of the lateral undulation and angle of intervertebral flexion is much greater during terrestrial locomotion than that of aquatic.

Musculoskeletal System

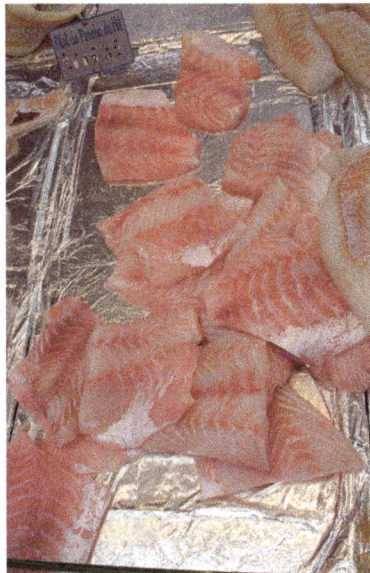
Perch filets showing myomere structure

Muscle Architecture

A typical characteristic of many animals that utilize undulatory locomotion is that they have segmented muscles, or blocks of myomeres, running from their head to tails which are separated by connective tissue called myosepta. In addition, some segmented muscle groups, such as the lateral hypaxial musculature in the salamander are oriented at an angle to the longitudinal direction. For these obliquely oriented fibers the strain in the longitudinal direction is greater than the strain in the muscle fiber direction leading to an architectural gear ratio greater than 1. A higher initial angle of orientation and more dorsoventral bulging produces a faster muscle contraction but results in a lower amount of force production. It is hypothesized that animals employ a variable gearing mechanism that allows self-regulation of force and velocity to meet the mechanical demands of the contraction. When a pennate muscle is subjected to a low force, resistance to width changes in the muscle cause it to rotate which consequently produce a higher architectural gear ratio (AGR) (high velocity). However, when subject to a high force, the perpendicular fiber force component overcomes the resistance to width changes and the muscle compresses producing a lower AGR (capable of maintaining a higher force output).

Most fishes bend as a simple, homogenous beam during swimming via contractions of longitudinal red muscle fibers and obliquely oriented white muscle fibers within the segmented axial musculature. The fiber strain (εf) experienced by the longitudinal red muscle fibers is equivalent to the longitudinal strain (εx). The deeper white muscle fibers fishes show diversity in arrangement. These fibers are organized into cone-shaped structures and attach to connective tissue sheets known as myosepta; each fiber shows a characteristic dorsoventral (α) and mediolateral (φ) trajectory. The segmented architecture theory predicts that, $\varepsilon x > \varepsilon f$. This phenomenon results in an architectural gear ratio, determined as longitudinal strain divided by fiber strain ($\varepsilon x / \varepsilon f$), greater than one and longitudinal velocity amplification; furthermore, this emergent velocity amplification may be augmented by variable architectural gearing via mesolateral and dorsoventral shape changes, a pattern seen in pennate muscle contractions. A red-to-white gearing ratio (red εf / white εf) captures the combined effect of the longitudinal red muscle fiber and oblique white muscle fiber strains.

Simple bending behavior in homogenous beams suggests ε increases with distance from the neutral axis (z). This poses a problem to animals, such as fishes and salamanders, which undergo undulatory movement. Muscle fibers are constrained by the length-tension and force-velocity curves. Furthermore, it has been hypothesized that muscle fibers recruited for a particular task must operate within an optimal range of strains (ε) and contractile velocities to generate peak force and power respectively. Non-uniform ε generation during undulatory movement would force differing muscle fibers recruited for the same task to operate on differing portions of the length-tension and force-velocity curves; performance would not be optimal. Alexander predicted that the dorsoventral (α) and mediolateral (φ) orientation of the white fibers of the fish axial musculature may allow more uniform strain across varying mesolateral fiber distances. Unfortunately, the white muscle fiber musculature of fishes is too complex to study uniform strain generation; however, Azizi *et al.* studied this phenomenon using a simplified salamander model.

Siren lacertian, an aquatic salamander, utilizes swimming motions similar to the aforementioned fishes yet contains hypaxial muscle fibers (which generate bending) characterized by a simpler organization. The hypaxial muscle fibers of S. lacertian are obliquely oriented, but have a near zero mediolateral (φ) trajectory and a constant doroslateral (α) trajectory within each segment. Therefore, the effect of doroslateral (α) trajectory and the distance between a given hypaxial muscle layer and the neutral axis of bending (z) on muscle fiber strain (ε) can be studied.

Azizi *et al.* found that longitudinal contractions of the constant volume hypaxial muscles were compensated by an increase in the dorsoventral dimensions. Bulging was accompanied by fiber rotation as well as an increase in both α hypaxial fiber trajectory and architectural gear ratio (AGR), a phenomenon also seen in pennate muscle contractions. Azizi *et al.* constructed a mathematical model to predict the final hypaxial fiber angle, AGR and dorsoventral height, where: λx = longitudinal extension ratio of the segment (portortion of final longitudinal length after contraction to initial longitudinal length), β = final fiber angle, γ = initial fiber angle, f = initial fiber length, and εx and εf = longitudinal and fiber strain respectively.

- $\lambda x = \varepsilon x + 1$

- $\lambda f = \varepsilon f + 1$

- $\varepsilon x = \Delta L / L_{inital}$

- Architectural gear ratio = $\varepsilon x / \varepsilon f = [\lambda f(\cos \beta / \cos \gamma) - 1)] / (\lambda f - 1)$

- $\beta = \sin^{-1} (y_2 / \lambda ff)$

This relationship shows that AGR increase with an increase in fiber angle from γ to β. In addition, final fiber angle (β) increases with dorsolateral bulging (y) and fiber contraction, but decreases as a function of initial fiber length.

The application of the latter conclusions can be seen in S. lacertian. This organism undulates as a homogenous beam (just as in fishes) during swimming; thus the distance of a muscle fiber from the neutral axis (z) during bending must be greater for external oblique muscle layers (EO) than internal oblique muscle layers (IO). The relationship between the strains (ε) experienced by the EO and IO and their respective z values is given by the following equation: where εEO and εIO = strain of the external and internal oblique muscle layers, and zEO and zIO = distance of the external and internal oblique muscle layers respectively from the neutral axis.

$$\varepsilon EO = \varepsilon IO (zEO / zIO)$$

Via this equation, we see that z is directly proportional to ε; the strain experienced by the EO exceeds that of the IO. Azizi *et al.* discovered that the initial hypaxial fiber α trajectory in the EO is greater than that of the IO. Because initial α trajectory is proportional to the AGR, the EO contracts with a greater AGR than the IO. The resulting velocity amplification allows both layers of muscles to operate at similar strains and shortening velocities; this enables the EO and IO to function on comparable portions of the length-tension and force-velocity curves. Muscles recruited for a similar task ought to operate at similar strains and velocities to maximize force and power output. Therefore, variability in AGR within the hypaxial musculature of the Siren lacertian counteracts varying mesolateral fiber distances and optimizes performance. Azizi *et al.* termed this phenomenon as fiber strain homogeneity in segmented musculature.

Muscle Activity

In addition to a rostral to caudal kinematic wave that travels down the animals body during undulatory locomotion, there is also a corresponding wave of muscle activation that travels in the rostro-caudal direction. However, while this pattern is characteristic of undulatory locomotion, it too can vary with environment.

American Eel

Aquatic Locomotion: Electromyogram (EMG) recordings of the American eel reveal a similar pattern of muscle activation during aquatic movement as that of fish. At slow speeds only the most posterior end of the eel's muscles are activated with more anterior muscle recruited at higher speeds. As in many other animals, the muscles activate late in the lengthening phase of the muscle strain cycle, just prior to muscle shortening which is a pattern believed to maximize work output from the muscle.

Terrestrial Locomotion: EMG recordings show a longer absolute duration and duty cycle of muscle activity during locomotion on land. Also, the absolute intensity is much higher while on land which is expected from the increase in gravitational forces acting on the animal. However, the intensity level decreases more posteriorly along the length of the eel's body. Also, the timing of muscle activation shifts to later in the strain cycle of muscle shortening.

Energetics

Animals with elongated bodies and reduced or no legs have evolved differently from their limbed relatives. In the past, some have speculated that this evolution was due to a lower energetic cost associated with limbless locomotion. The biomechanical arguments used to support this rationale include that (1) there is no cost associatied with the vertical displacement of the center of mass typically found with limbed animals, (2) there is no cost associated with accelerating or decelerating limbs, and (3) there is a lower cost for supporting the body. This hypothesis has been studied further by examining the oxygen consumption rates in the snake during different modes of locomotion: lateral undulation, concertina, and sidewinding. The *net cost of transport* (NCT), which indicates the amount of energy required to move a unit of mass a given distance, for a snake moving with a lateral undulatory gait is identical to that of a limbed lizard with the same mass. However, a snake utilizing concertina locomotion produces a much higher net cost of transport, while sidewinding actually produces a lower net cost of transport. Therefore, the different modes of locomotion are of primary importance when determining energetic cost. The reason that lateral undulation has the same energetic efficiency as limbed animals and not less, as hypothesized earlier, might be due to the additional biomechanical cost associated with this type of movement due to the force needed to bend the body laterally, push its sides against a vertical surface, and overcome sliding friction.

Neuromuscular System

Intersegmental Coordination

Wavelike motor pattern typically arise from a series of coupled segmental oscillator. Each segmental oscillator is capable of producing a rhythmic motor output in the absence of sensory feedback. One such example is the half center oscillator which consist of two neurons that are mutually inhibitory and produce activity 180 degrees out of phase. The phase relationships between these oscillators are established by the emergent properties of the oscillators and the coupling between them. Forward swimming can by accomplished by a series of coupled oscillators in which the anterior oscillators have a shorter endogenous frequency than the posterior oscillators. In this case, all oscillators will be driven at the same period but the anterior oscillators will lead in phase. In addition, the phase relations can be established by asymmetries in the couplings between oscillators or by sensory feedback mechanisms.

- Leech

The leech moves by producing dorsoventral undulations. The phase lags between body segments is about 20 degrees and independent of cycle period. Thus, both hemisegments of the oscillator fire synchronously to produce a contraction. Only the ganglia rostral to the midpoint are capable of producing oscillation individually. There is U-shaped gradient in endogenous segment oscillation as well with the highest oscillations frequencies occurring near the middle of the animal. Although the couplings between neurons spans six segments in both the anterior and posterior direction, there are asymmetries between the various interconnections because the oscillators are active at three different phases. Those that are active in the 0 degree phase project only in the descending direction while those projecting in the ascending direction are active at 120 degrees or 240 degrees. In addition, sensory feedback from the environment may contribute to resultant phase lag.

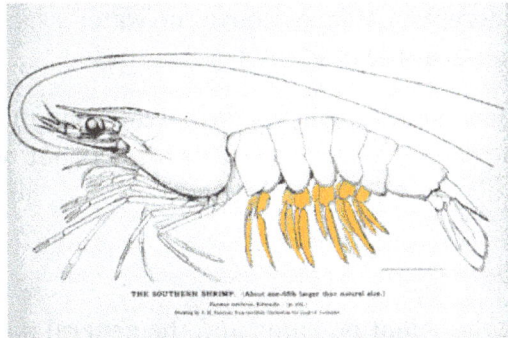

THE SOUTHERN SHRIMP. (About one-fifth larger than natural size.)

Pleopods

- Lamprey

The lamprey moves using lateral undulation and consequently left and right motor hemisegments are active 180 degrees out of phase. Also, it has been found that the endogenous frequency of the more anterior oscillators is higher than that of the more posterior ganglia. In addition, inhibitory interneurons in the lamprey project 14-20 segments caudally but have short rostral projections. Sensory feedback may be important for appropriately responding to perturbations, but seems to be less important for maintaince of appropriate phase relations.

Robotics

Based on biologically hypothesized connections of the central pattern generator in the salamander, a robotic system has been created which exhibits the same characteristics of the actual animal. Electrophysiology studies have shown that stimulation of the mesencephalic locomotor region (MLR) located in the brain of the salamander produce different gaits, swimming or walking, depending on intensity level. Similarly, the CPG model in the robot can exhibit walking at low levels of tonic drive and swimming at high levels of tonic drive. The model is based on the four assumptions that:

- Tonic stimulation of the body CPG produces spontaneous traveling waves. When the limb CPG is activated it overrides the body CPG.

- The strength of the coupling from the limb to the body CPG is stronger than that from body to limb.

- Limb oscillators saturate and stop oscillating at higher tonic drives.

- Limb oscillators have lower intrinsic frequencies than body CPGs at the same tonic drive.

This model encompasses the basic features of salamander locomotion.

Flying Fish

The Exocoetidae are a family of marine fish in the order Beloniformes class Actinopterygii. Fish of this family are known as flying fish. About 64 species are grouped in seven to nine genera. Flying fish can make powerful, self-propelled leaps out of water into air, where their long, wing-like fins enable gliding flight for considerable distances above the water's surface. This uncommon ability is a natural defence mechanism to evade predators.

The oldest known fossil of a flying or gliding fish, *Potanichthys xingyiensis*, dates back to the Middle Triassic, 235–242 million years ago. However, this fossil is not related to modern flying fish, which evolved independently about 66 million years ago.

Etymology

The term Exocoetidae is both the scientific name and the general name in Latin for a flying fish. The suffix *-idae*, common for indicating a family, follows the root of the Latin word *exocoetus*, a transliteration of the Ancient Greek name. This means literally sleeping outside, from outside and bed, resting place, so named as flying fish were believed to leave the water to sleep on the shore.

A different etymological approach, more realistic, is that of Ancient Greek outside with hull, which means not submerged in the water.

Distribution and Description

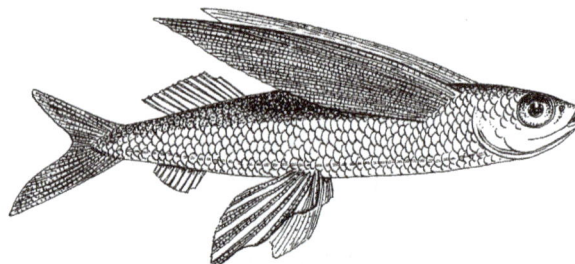

Flying fish

Flying fish live in all of the oceans, particularly in tropical and warm subtropical waters. They are commonly found in the epipelagic zone. This area is the top layer of the ocean that extends 200 meters from the surface down. It is often known as the "sunlight zone" because it's where most of the visible light exists. Nearly all primary production, or photosynthesis, happens in this zone.

Therefore, the vast majority of plants and animals inhabit this area and can vary from plankton to the sharks. Although the epipelagic zone is an exceptional area for variety in life, it too has its drawbacks. Due to the vast variety of organisms it holds, there is high number of prey and predation relationships. Small organisms such as the flying fish are targets for larger organisms. They especially have a hard time escaping predators and surviving until they can reproduce, resulting in them having a lower fitness. Along with relationship difficulties, abiotic factors also play a part. Harsh ocean currents make it extremely difficult for small fish to survive in this habitat. Research suggests that difficult environmental factors in the flying fish's habitat have led to the evolution of modified fins. As a result, flying fish have undergone natural selection in which species gain unique traits to better adapt to their environments. By becoming airborne, flying fish evade their predators and environment. This increase of speed and maneuverability is a direct advantage to flying fish, and has given them leverage when compared to other species in their environment.

Flying fish taking off

Research has shown that the flying fish has undergone morphological changes throughout its history, the first of which is fully broadened neural arches. Neural arches act as insertion sites for muscles, connective tissues, and ligaments in a fish's skeleton. Fully broadened neural arches act as more stable and sturdier sites for these connections, creating a strong link between the vertebral column and cranium. This ultimately allows a rigid and sturdy vertebral column (body) that is beneficial in flight. Having a rigid body during glided flight gives the flying fish aerodynamic advantages, increasing its speed and improving its aim. Furthermore, flying fish have developed vertebral columns and ossified caudal complexes. These features provide the majority of strength to the flying fish, allowing them to physically lift their body out of water and glide remarkable distances. These additions also reduce the flexibility of the flying fish, allowing them to perform powerful leaps without weakening midair. At the end of a glide, it folds its pectoral fins to re-enter the sea, or drops its tail into the water to push against the water to lift itself for another glide, possibly changing direction. The curved profile of the "wing" is comparable to the aerodynamic shape of a bird wing. The fish is able to increase its time in the air by flying straight into or at an angle to the direction of updrafts created by a combination of air and ocean currents.

Genus *Exocoetus* has one pair of fins and a streamlined body to optimize for speed, while *Cypselurus* has a flattened body and two pairs of fins, which maximize its time in the air. From 1900 to the 1930s, flying fish were studied as possible models used to develop airplanes.

Exocoetidae feed mainly on plankton. Predators include dolphins, tuna, marlin, birds, squids, and porpoises.

Flight Measurements

In May 2008, a Japanese television crew (NHK) filmed a flying fish (dubbed "Icarfish") off the coast of Yakushima Island, Japan. The fish spent 45 seconds in flight. The previous record was 42 seconds.

The flights of flying fish are typically around 50 meters (160 ft), though they can use updrafts at the leading edge of waves to cover distances of up to 400 m (1,300 ft). They can travel at speeds of more than 70 km/h (43 mph). Maximum altitude is 6 m (20 ft) above the surface of the sea. Some accounts (e.g. Kon-tiki by Thor Heyerdal) have them landing on ships' decks.

Fishery and Cuisine

Flying fish for sale in local fish market of Saint Martin's Island, Bangladesh

Flying fish are commercially fished in Japan, Vietnam, and China by gillnetting, and in Indonesia and India by dipnetting. Often in Japanese cuisine, the fish is preserved by drying. The roe of *Cheilopogon agoo*, or Japanese flying fish, is used to make some types of sushi, and is known as *tobiko*. It is also a staple in the diet of the Tao people of Orchid Island, Taiwan. Flying fish is part of the national dish of Barbados, *cou-cou* and flying fish.

In the Solomon Islands, the fish are caught while they are flying, using nets held from outrigger canoes. They are attracted to the light of torches. Fishing is done only when there is no moonlight.

Importance

Barbados

Barbados is known as "the land of the flying fish", and the fish is one of the national symbols of the country. Once abundant, it migrated between the warm, coral-filled Atlantic Ocean surrounding the island of Barbados and the plankton-rich outflows of the Orinoco River in Venezuela.

Just after the completion of the Bridgetown Harbor / Deep Water Harbor in Bridgetown, Barbados saw an increase of ship visits, linking the island to the world. The overall health of the coral reefs surrounding Barbados suffered due to ship-based pollution. Additionally, Barbadian overfishing pushed them closer to the Orinoco delta, no longer returning to Barbados in large numbers. Today, the flying fish only migrate as far north as Tobago, around 120 nmi (220 km; 140 mi) southwest of Barbados. Despite the change, flying fish remain a coveted delicacy.

Many aspects of Barbadian culture center around the flying fish: it is depicted on coins, as sculptures in fountains, in artwork, and as part of the official logo of the Barbados Tourism Authority. Additionally, the Barbadian coat of arms features a pelican and dolphin fish on either side of the shield, but the dolphin resembles a flying fish. Furthermore, actual artistic renditions and holograms of the flying fish are also present within the Barbadian passport.

Maritime Disputes

In recent times, flying fish have also been gaining in popularity in other islands, fueling several maritime disputes. In 2006, the council of the United Nations Convention on the Law of the Sea fixed the maritime boundaries between Barbados and Trinidad and Tobago over the flying fish dispute, which gradually raised tensions between the neighbours. The ruling stated both countries must preserve stocks for the future. Barbadian fishers still follow the flying fish southward. Flying fish remain an important part of Barbados' main national dish.

Walking Fish

A walking fish, or ambulatory fish, is a fish that is able to travel over land for extended periods of time. Some other modes of non-standard fish locomotion include "walking" along the sea floor, for example, in handfish or frogfish.

Types

Most commonly, walking fish are amphibious fish. Able to spend longer times out of water, these fish may use a number of means of locomotion, including springing, snake-like lateral undulation, and tripod-like walking. The mudskippers are probably the best land-adapted of contemporary fish and are able to spend days moving about out of water and can even climb mangroves, although to only modest heights. The climbing gourami is often specifically referred to as a "walking fish", although it does not actually "walk", but rather moves in a jerky way by supporting itself on the extended edges of its gill plates and pushing itself by its fins and tail. Some reports indicate that it can also climb trees.

The epaulette shark (*Hemiscyllium ocellatum*) tends to live in shallow waters where swimming is difficult, and can often be seen walking over rocks and sand by using its muscular pectoral fins. It lives in areas of great variation in water depth, usually where the tide falls below its location. If it finds itself out of water, it can survive for several hours, and is capable of walking over land to get to water. This means that it is easily observed by beachgoers in its natural range.

There are a number of fish that are less adept at actual walking, such as the walking catfish. Despite being known for "walking on land", this fish usually wriggles and may use its pectoral fins to aid in its movement. Walking catfish have a respiratory system that allows them to live out of water for several days. Some are invasive species, for example, the Northern snakehead in the U.S. Polypterids have rudimentary lungs and can also move about on land, though rather clumsily. The mangrove rivulus can survive for months out of water and can move to places like hollow logs.

Some species of fish can "walk" along the sea floor but not on land. One such animal is the flying gurnard (it does not actually fly, and should not be confused with flying fish). The batfishes of the Ogcocephalidae family (not to be confused with Batfish of Ephippidae) are also capable of walking along the sea floor. *Bathypterois grallator*, also known as a "tripodfish", stands on three fins on the bottom of the ocean and hunts for food. The African lungfish (P. annectens) can use its fins to "walk" along the bottom of its tank in a manner similar to the way amphibians and land vertebrates use their limbs on land.

Evolutionary Link

Tiktaalik (reconstruction)

In modern fish, the "walking" ability differs from that of tetrapods (four-limbed animals). The theory of evolution suggests that life originated in the oceans and later moved onto land, and paleontologists have long been looking for a missing evolutionary link between ocean-living and land-living animals. In 2006, a fossil *Tiktaalik roseae* was found which has many features of wrist, elbow, and neck that are akin to those of tetrapods. It belongs to a group of lobe-finned fish called *Rhipidistia*, which according to some theories, were the ancestors of all tetrapods.

References

- Sfakiotakis, M.; Lane, D. M.; Davies, J. B. C. (1999). "Review of Fish Swimming Modes for Aquatic Locomotion" (PDF). IEEE Journal of Oceanic Engineering. 24 (2)

- Allen, Gerry (1999). Marine Fishes of Southeast Asia: A Field Guide for Anglers and Divers. Tuttle Publishing. p. 56. ISBN 978-1-4629-1707-5. many have a bony, sharp tail and are equally adept at burrowing forward or backward

- Matthew J. and George V. Lauder. (2006) Ontogeny of Form and Function: Locomotor Morphology and Drag in Zebrafish (Danio rerio). "Journal of Morphology." 267,1099-1109

- Ord, T. J.; Summers, T. C.; Noble, M. M.; Fulton, C. J.; McPeek, M. A. (2017-03-02). "Ecological Release from Aquatic Predation Is Associated with the Emergence of Marine Blenny Fishes onto Land". The American Naturalist: 000–000. doi:10.1086/691155

- Brainerd, E. L.; Azizi, E. (2005). "Muscle Fiber Angle, Segment Bulging and Architectural Gear Ratio in Segmented Musculature". Journal of Experimental Biology. 208 (17): 3249–3261. PMID 16109887. doi:10.1242/jeb.01770

- Olsen, Stanley J. (1968). Fish, Amphibian and Reptile Remains from Archaeological Sites. Acme Bookbinding. p. 4. ISBN 0-87365-163-4

- Guo, Z. V.; Mahadeven, L. (2008). "Limbless undulatory propulsion on land". PNAS. 105 (9): 3179–3184. PMC 2265148. PMID 18308928. doi:10.1073/pnas.0705442105

- Brainerd, E. L.; Azizi, E. (2007). "Architectural Gear Ratio and Muscle Fiber Strain Homogeneity in Segmented Musculature". Journal of Experimental Zoology. 307 (A): 145–155. doi:10.1002/jez.a.358

- Joseph Banks (1997). The Endeavour Journal of Sir Joseph Banks 1768–1771 (PDF). University of Sydney Library. Retrieved July 16, 2009

- Secor, S. M.; Jayne, B. C.; Bennett, A. F. "Locomotor Performance and energetic Cost of Sidewinging by the Snake Crotalus Cerastes". Journal of Experimental Biology. 163 (1): 1–14

- Gans, C. (1975). "Tetrapod Limblessness: Evolution and Functional Corollaries". Am. Zool. 15 (2): 455–461. doi:10.1093/icb/15.2.455

- Hill, Andrew A. V.; Masino, Mark A.; Calabrese, Ronald L. (2003). "Intersegmental Coordination of Rhythmic Motor Patterns". Journal of Neurophysiology. 90 (2): 531–538. PMID 12904484. doi:10.1152/jn.00338.2003

- Hanken, James; Hall, Brian K. (1993). The Skull, Volume 2: Patterns of Structural and Systematic Diversity. University of Chicago Press. p. 209. ISBN 0-226-31570-3

- Fish, F. E. (1990). Wing Design And Scaling of Flying Fish With Regard To Flight Performance. Journal of Zoology, 221(3), 391-403

- Ijspeert, A. J. (2001). "A Connectionist Central Pattern Generator for the Aquatic and Terrestrial Gaits of a Simulated Salamander". Biological Cybernetics. 84 (5): 331–348. doi:10.1007/s004220000211

- C. M. Pace and A. C. Gibb (July 15, 2009). "Mudskipper pectoral fin kinematics in aquatic and terrestrial environments" (PDF). The Journal of Experimental Biology. 212: 2279–2286. PMID 19561218. doi:10.1242/jeb.029041

5

Fish Reproduction

The method of reproduction in fishes varies. The five modes of reproduction in fishes are ovuliparity, oviparity, ovoviviparity, hemotrophic and histotrophic viviparity. This chapter has been carefully written to provide an easy understanding of the varied types of fish reproduction.

Fish Reproduction

Fish reproductive organs include testes and ovaries. In most species, gonads are paired organs of similar size, which can be partially or totally fused. There may also be a range of secondary organs that increase reproductive fitness. The genital papilla is a small, fleshy tube behind the anus in some fishes, from which the sperm or eggs are released; the sex of a fish often can be determined by the shape of its papilla.

Anatomy

Testes

Most male fish have two testes of similar size. In the case of sharks, the testes on the right side is usually larger. The primitive jawless fish have only a single testis, located in the midline of the body, although even this forms from the fusion of paired structures in the embryo.

Under a tough membranous shell, the tunica albuginea, the testis of some teleost fish, contains very fine coiled tubes called seminiferous tubules. The tubules are lined with a layer of cells (germ cells) that from puberty into old age, develop into sperm cells (also known as spermatozoa or male gametes). The developing sperm travel through the seminiferous tubules to the rete testis located in the mediastinum testis, to the efferent ducts, and then to the epididymis where newly created sperm cells mature. The sperm move into the vas deferens, and are eventually expelled through the urethra and out of the urethral orifice through muscular contractions.

However, most fish do not possess seminiferous tubules. Instead, the sperm are produced in spherical structures called *sperm ampullae*. These are seasonal structures, releasing their contents during the breeding season, and then being reabsorbed by the body. Before the next breeding season, new sperm ampullae begin to form and ripen. The ampullae are otherwise essentially identical to the seminiferous tubules in higher vertebrates, including the same range of cell types.

In terms of spermatogonia distribution, the structure of teleosts testes has two types: in the most common, spermatogonia occur all along the seminiferous tubules, while in Atherinomorph fish they are confined to the distal portion of these structures. Fish can present cystic or semi-cystic spermatogenesis in relation to the release phase of germ cells in cysts to the seminiferous tubules lumen.

Ovaries

Many of the features found in ovaries are common to all vertebrates, including the presence of follicular cells and tunica albuginea There may be hundreds or even millions of fertile eggs present in the ovary of a fish at any given time. Fresh eggs may be developing from the germinal epithelium throughout life. Corpora lutea are found only in mammals, and in some elasmobranch fish; in other species, the remnants of the follicle are quickly resorbed by the ovary. The ovary of teleosts is often contains a hollow, lymph-filled space which opens into the oviduct, and into which the eggs are shed. Most normal female fish have two ovaries. In some elasmobranchs, only the right ovary develops fully. In the primitive jawless fish, and some teleosts, there is only one ovary, formed by the fusion of the paired organs in the embryo.

Fish ovaries may be of three types: gymnovarian, secondary gymnovarian or cystovarian. In the first type, the oocytes are released directly into the coelomic cavity and then enter the ostium, then through the oviduct and are eliminated. Secondary gymnovarian ovaries shed ova into the coelom from which they go directly into the oviduct. In the third type, the oocytes are conveyed to the exterior through the oviduct. Gymnovaries are the primitive condition found in lungfish, sturgeon, and bowfin. Cystovaries characterize most teleosts, where the ovary lumen has continuity with the oviduct. Secondary gymnovaries are found in salmonids and a few other teleosts.

Eggs

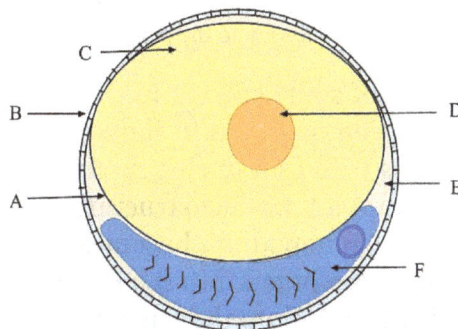

Diagram of a fish egg: A. vitelline membrane B. chorion C. yolk D. oil globule E. perivitelline space F. embryo

The eggs of fish and amphibians are jellylike. Cartilagenous fish (sharks, skates, rays, chimaeras) eggs are fertilized internally and exhibit a wide variety of both internal and external embryonic development. Most fish species spawn eggs that are fertilized externally, typically with the male inseminating the eggs after the female lays them. These eggs do not have a shell and would dry out in the air. Even air-breathing amphibians lay their eggs in water, or in protective foam as with the Coast foam-nest treefrog, *Chiromantis xerampelina*.

Intromittent Organs

Male cartilaginous fishes (sharks and rays), as well as the males of some live-bearing ray finned fishes, have fins that have been modified to function as intromittent organs, reproductive appendages which allow internal fertilization. In ray finned fish they are called *gonopodiums* or *andropodiums*, and in cartilaginous fish they are called *claspers*.

This male mosquitofish has a gonopodium, an anal fin which functions as an intromittent organ

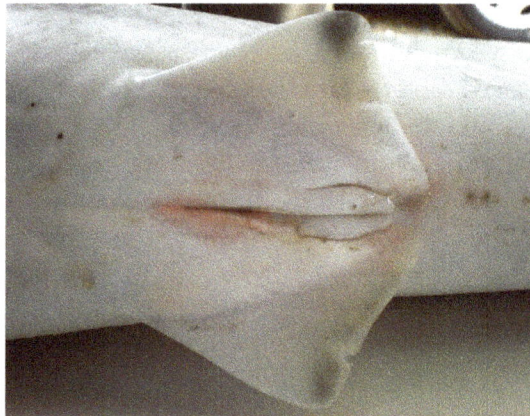

This young male spinner shark has claspers, a modification to the pelvic fins which also function as intromittent organs

Gonopodia are found on the males of some species in the Anablepidae and Poeciliidae families. They are anal fins that have been modified to function as movable intromittent organs and are used to impregnate females with milt during mating. The third, fourth and fifth rays of the male's anal fin are formed into a tube-like structure in which the sperm of the fish is ejected. When ready for mating, the gonopodium becomes erect and points forward towards the female. The male shortly inserts the organ into the sex opening of the female, with hook-like adaptations that allow the fish to grip onto the female to ensure impregnation. If a female remains stationary and her partner contacts her vent with his gonopodium, she is fertilized. The sperm is preserved in the female's oviduct. This allows females to fertilize themselves at any time without further assistance from males. In some species, the gonopodium may be half the total body length. Occasionally the fin is too long to be used, as in the "lyretail" breeds of *Xiphophorus helleri*. Hormone treated females may develop gonopodia. These are useless for breeding.

Similar organs with similar characteristics are found in other fishes, for example the *andropodium* in the *Hemirhamphodon* or in the Goodeidae.

Claspers are found on the males of cartilaginous fishes. They are the posterior part of the pelvic fins that have also been modified to function as intromittent organs, and are used to channel semen into the female's cloaca during copulation. The act of mating in sharks usually includes raising one of the claspers to allow water into a siphon through a specific orifice. The clasper is then inserted into the cloaca, where it opens like an umbrella to anchor its position. The siphon then begins to contract expelling water and sperm.

Physiology

Oogonia development in teleosts fish varies according to the group, and the determination of oogenesis dynamics allows the understanding of maturation and fertilisation processes. Changes in the nucleus, ooplasm, and the surrounding layers characterize the oocyte maturation process.

Postovulatory follicles are structures formed after oocyte release; they do not have endocrine function, present a wide irregular lumen, and are rapidly reabsorbed in a process involving the apoptosis of follicular cells. A degenerative process called follicular atresia reabsorbs vitellogenic oocytes not spawned. This process can also occur, but less frequently, in oocytes in other development stages.

Some fish are hermaphrodites, having both testes and ovaries either at different phases in their life cycle or, as in hamlets, have them simultaneously.

Reproductive Strategies

In fish, fertilisation of eggs can be either external or internal. In many species of fish, fins have been modified to allow Internal fertilisation. Similarly, development of the embryo can be external or internal, although some species show a change between the two at various stages of embryo development. Thierry Lodé described reproductive strategies in terms of the development of the zygote and the interrelationship with the parents; there are five classifications - ovuliparity, oviparity, ovo-viviparity, histotrophic viviparity and hemotrophic viviparity.

Ovuliparity

Ovuliparity means the female lays unfertilised eggs (ova), which must then be externally fertilised. Examples of ovuliparous fish include salmon, goldfish, cichlids, tuna and eels. In the majority of these species, fertilisation takes place outside the mother's body, with the male and female fish shedding their gametes into the surrounding water.

Oviparity

Oviparity is where fertilisation occurs internally and so the female sheds zygotes (or newly developing embryos) into the water, often with important outer tissues added. Over 97% of all known fish are oviparous, In oviparous fish, internal fertilisation requires the male to use some sort of intromittent organ to deliver sperm into the genital opening of the female. Examples include the oviparous sharks, such as the horn shark, and oviparous rays, such as skates. In these cases, the male is equipped with a pair of modified pelvic fins known as claspers.

Marine fish can produce high numbers of eggs which are often released into the open water column. The eggs have an average diameter of 1 millimetre (0.039 in). The eggs are generally surrounded by the extraembryonic membranes but do not develop a shell, hard or soft, around these membranes. Some fish have thick, leathery coats, especially if they must withstand physical force or desiccation. These type of eggs can also be very small and fragile.

The newly hatched young of oviparous fish are called larvae. They are usually poorly formed, carry a large yolk sac (for nourishment) and are very different in appearance from juvenile and adult specimens. The larval period in oviparous fish is relatively short (usually only several weeks), and larvae rapidly grow and change appearance and structure (a process termed metamorphosis) to become juveniles. During this transition larvae must switch from their yolk sac to feeding on zooplankton prey, a process which depends on typically inadequate zooplankton density, starving many larvae.

Ovoviviparity

In ovoviviparous fish the eggs develop inside the mother's body after internal fertilisation but receive little or no nourishment directly from the mother, depending instead on a food reserve inside the egg, the yolk. Each embryo develops in its own egg. Familiar examples of ovoviviparous fish include guppies, angel sharks, and coelacanths.

Viviparity

There are two types of viviparity, differentiated by how the offspring gain their nutrients.

- Histotrophic (tissue eating) viviparity means embryos develop in the female's oviducts but obtain nutrients by consuming other tissues, such as ova (oophagy) or zygotes. This has been observed primarily among sharks such as the shortfin mako and porbeagle, but is known for a few bony fish as well such as the halfbeak *Nomorhamphus ebrardtii*. An unusual mode of vivipary is adelphophagy or intrauterine cannibalism, in which the largest embryos eat weaker, smaller unborn siblings. This is most commonly found among sharks such as the grey nurse shark, but has also been reported for *Nomorhamphus ebrardtii*.

- Hemotrophic (blood eating) viviparity means embryos develop in the female's (or male's) oviduct and nutrients are provided directly by the parent, typically via a structure similar to, or analogous to the placenta seen in mammals. Examples of hemotrophic fish include the surfperches, splitfins, lemon shark, seahorses and pipefish.

Aquarists commonly refer to ovoviviparous and viviparous fish as livebearers.

Hermaphroditism

Female groupers change their sex to male if no male is available

An anemone fish couple guarding their anemone. If the female dies,
a juvenile male moves in, and the resident male changes sex.

Parthenogenesis was first described among vertebrates in the Amazon molly

Hermaphroditism occurs when a given individual in a species possesses both male and female reproductive organs, or can alternate between possessing first one, and then the other. Hermaphroditism is common in invertebrates but rare in vertebrates. It can be contrasted with gonochorism, where each individual in a species is either male or female, and remains that way throughout their lives. Most fish are gonochorists, but hermaphroditism is known to occur in 14 families of teleost fishes.

Usually hermaphrodites are *sequential*, meaning they can switch sex, usually from female to male (protogyny). This can happen if a dominant male is removed from a group of females. The largest female in the harem can switch sex over a few days and replace the dominant male. This is found amongst coral reef fishes such as groupers, parrotfishes and wrasses. It is less common for a male to switch to a female (protandry). As an example, most wrasses are protogynous hermaphrodites within a haremic mating system. Hermaphroditism allows for complex mating systems. Wrasses exhibit three different mating systems: polygynous, lek-like, and promiscuous mating systems. Group spawning and pair spawning occur within mating systems. The type of spawning that occurs depends on male body size. Labroids typically exhibit broadcast spawning, releasing high amounts of planktonic eggs, which are broadcast by tidal currents; adult wrasses have no interaction with offspring. Wrasse of a particular subgroup of the Labridae family Labrini do not exhibit broadcast spawning.

Less commonly hermaphrodites can be *synchronous*, meaning they simultaneously possess both ovaries and testicles and can function as either sex at any one time. Black hamlets "take turns releasing sperm and eggs during spawning. Because such egg trading is advantageous to both individuals, hamlets are typically monogamous for short periods of time—an unusual situation in fishes." The sex of many fishes is not fixed, but can change with physical and social changes to the environment where the fish lives.

Particularly among fishes, hermaphroditism can pay off in situations where one sex is more likely to survive and reproduce, perhaps because it is larger. Anemone fishes are sequential hermaphrodites which are born as males, and become females only when they are mature. Anemone fishes live together monogamously in a anemone, protected by the anemone stings. The males do not have to compete with other males, and female anemone fish are typically larger. When a female dies a juvenile (male) anemone fish moves in, and "the resident male then turns into a female and reproductive advantages of the large female—small male combination continue". In other fishes sex changes are reversible. For example, if some gobies are grouped by sex (male or female), some will switch sex.

The mangrove rivulus *Kryptolebias marmoratus* produces both eggs and sperm by meiosis and routinely reproduces by self-fertilization. Each individual hermaphrodite normally fertilizes itself when an egg and sperm that it has produced by an internal organ unite inside the fish's body. In nature, this mode of reproduction can yield highly homozygous lines composed of individuals so genetically uniform as to be, in effect, identical to one another. The capacity for selfing in these fishes has apparently persisted for at least several hundred thousand years.

Although inbreeding, especially in the extreme form of self-fertilization, is ordinarily regarded as detrimental because it leads to expression of deleterious recessive alleles, self-fertilization does provide the benefit of *fertilization assurance* (reproductive assurance) at each generation.

Sexual Parasitism

Some anglerfish, like those of the deep sea ceratioid group, employ an unusual mating method. Because individuals are locally rare, encounters are also very rare. Therefore, finding a mate is problematic. When scientists first started capturing ceratioid anglerfish, they noticed that all of the specimens were female. These individuals were a few centimetres in size and almost all of them had what appeared to be parasites attached to them. It turned out that these "parasites" were highly reduced male ceratioids. This indicates the anglerfish use a polyandrous mating system.

The methods by which the anglerfish locate mates are variable. Some species have minute eyes unfit for identifying females visually, while others have underdeveloped nostrils, making it unlikely that they effectively find females using olfaction. When a male finds a female, he bites into her skin, and releases an enzyme that digests the skin of his mouth and her body, fusing the pair down to the blood-vessel level. The male becomes dependent on the female host for survival by receiving nutrients via their shared circulatory system, and provides sperm to the female in return. After fusing, males increase in volume and become much larger relative to free-living males of the species. They live and remain reproductively functional as long as the female lives, and can take part in multiple spawnings. This extreme sexual dimorphism ensures, when the female is ready to spawn, she has a mate immediately available. Multiple males can be incorporated into a single individual female with up to eight males in some species, though some taxa appear to have a one male per female rule.

One explanation for the evolution of sexual parasitism is that the relative low density of females in deep-sea environments leaves little opportunity for mate choice among anglerfish. Females remain large to accommodate fecundity, as is evidenced by their large ovaries and eggs. Males would be expected to shrink to reduce metabolic costs in resource-poor environments and would develop highly specialized female-finding abilities. If a male manages to find a female parasitic attachment, then it is ultimately more likely to improve lifetime fitness relative to free living, particularly when the prospect of finding future mates is poor. An additional advantage to parasitism is that the male's sperm can be used in multiple fertilizations, as he stays always available to the female for mating. Higher densities of male-female encounters might correlate with species that demonstrate facultative parasitism or simply use a more traditional temporary contact mating.

Parthenogenesis

Parthenogenesis is a form of asexual reproduction in which growth and development of embryos occur without fertilization. In animals, parthenogenesis means development of an embryo from an unfertilized egg cell. The first all-female (unisexual) reproduction in vertebrates was described in the *Amazon molly* in 1932. Since then at least 50 species of unisexual vertebrate have been described, including at least 20 fish, 25 lizards, a single snake species, frogs, and salamanders. As with all types of asexual reproduction, there are both costs (low genetic diversity and therefore susceptibility to adverse mutations that might occur) and benefits (reproduction without the need for a male) associated with parthenogenesis.

Parthenogenesis in sharks has been confirmed in the bonnethead and zebra shark. Other, usually sexual species, may occasionally reproduce parthenogenetically, and the hammerhead and black-tip sharks are recent additions to the known list of facultative parthenogenetic vertebrates.

A special case of parthenogenesis is gynogenesis. In this type of reproduction, offspring are produced by the same mechanism as in parthenogenesis, however, the egg is stimulated to develop simply by the *presence* of sperm - the sperm cells do not contribute any genetic material to the offspring. Because gynogenetic species are all female, activation of their eggs requires mating with males of a closely related species for the needed stimulus. The Amazon molly, (pictured), reproduces by gynogenesis.

Others

The elkhorn sculpin (*Alcichthys elongatus*) is a marine teleost with a unique reproductive mode called "internal gametic association". Sperm are introduced into the ovary by copulation and then enter the micropylar canal of ovulated eggs in the ovarian cavity. However, actual sperm-egg fusion does not occur until the eggs have been released into sea water.

Inbreeding

Inbreeding Depression

The effect of inbreeding on reproductive behavior was studied in the poeciliid fish *Heterandria formosa*. One generation of full-sib mating was found to decrease reproductive performance and likely reproductive success of male progeny. Other traits that displayed inbreeding depression were offspring viability and maturation time of both males and females.

Exposure of zebra fish to a chemical environmental agent, analogous to that caused by anthropogenic pollution, amplified the effects of inbreeding on key reproductive traits. Embryo viability was significantly reduced in inbred exposed fish and there was a tendency for inbred males to sire fewer offspring.

The behaviors of juvenile Coho salmon with either low or medium inbreeding were compared in paired contests. Fish with low inbreeding showed almost twice the aggressive pursuit in defending territory than fish with medium inbreeding, and furthermore had a higher specific growth rate. A significant effect of inbreeding depression on juvenile survival was also found, but only in high-density competitive environments, suggesting that intra-specific competition can magnify the deleterious effects of inbreeding.

Inbreeding Avoidance

Inbreeding ordinarily has negative fitness consequences (inbreeding depression), and as a result species have evolved mechanisms to avoid inbreeding. Numerous inbreeding avoidance mechanisms operating prior to mating have been described. However, inbreeding avoidance mechanisms that operate subsequent to copulation are less well known. In guppies, a post-copulatory mechanism of inbreeding avoidance occurs based on competition between sperm of rival males for achieving fertilisation. In competitions between sperm from an unrelated male and from a full sibling male, a significant bias in paternity towards the unrelated male was observed.

Inbreeding depression is considered to be due largely to the expression of homozygous deleterious recessive mutations. Outcrossing between unrelated individuals results in the beneficial masking of deleterious recessive mutations in progeny.

Examples

Female goldfish sp awn (discharge) eggs into the water, encouraged by male goldfish who simultaneously discharge sperm which externally fertilizes the eggs

Within two or three days, the vulnerable goldfish eggs hatch into larvae, and rapidly develop into fry

Goldfish

Goldfish, like all cyprinids, are egg-layers. They usually start breeding after a significant temperature change, often in spring. Males chase females, prompting them to release their eggs by bumping and nudging them. As the female goldfish spawns her eggs, the male goldfish stays close behind fertilizing them. Their eggs are adhesive and attach to aquatic vegetation. The eggs hatch within 48 to 72 hours. Within a week or so, the fry begins to assume its final shape, although a year may pass before they develop a mature goldfish colour; until then they are a metallic brown like their wild ancestors. In their first weeks of life, the fry grow quickly—an adaptation born of the high risk of getting devoured by the adult goldfish.

Carp

A member of the Cyprinidae family, carp spawn in times between April and August, largely dependent upon the climate and conditions they live in. Oxygen levels of the water, availability of food, size of each fish, age, number of times the fish has spawned before and water temperature are all factors known to effect when and how many eggs each carp will spawn at any one time.

Siamese Fighting Fish

Prior to spawning, male Siamese fighting fish build bubble nests of varying sizes at the surface of the water. When a male becomes interested in a female, he will flare his gills, twist his body, and spread his fins. The female darkens in colour and curves her body back and forth. The act of spawning takes place in a "nuptial embrace" where the male wraps his body around the female, each embrace resulting in the release of 10-40 eggs until the female is exhausted of eggs. The male, from his side, releases milt into the water and fertilization takes place externally. During and after spawning, the male uses his mouth to retrieve sinking eggs and deposit them in the bubble nest (during mating the female sometimes assists her partner, but more often she will simply devour all the eggs that she manages to catch). Once the female has released all of her eggs, she is chased away from the male's territory, as it is likely that she'll eat the eggs due to hunger. The eggs then remain in the male's care. He keeps them in the bubble nest, making sure none fall to the bottom and repairing the nest as needed. Incubation lasts for 24–36 hours, and the newly hatched larvae remain in the nest for the next 2–3 days, until their yolk sacs are fully absorbed. Afterwards the fry leave the nest and the free-swimming stage begins.

| Siamese fighting fish build bubble nests of varying sizes. | A pair of Siamese fighting fish spawning under their bubble nest. | A 15-day-old free-swimming fry of a Siamese fighting fish |

Spawn (Biology)

The spawn (eggs) of a clownfish. The black spots are the developing eyes.

Spawn is the eggs and sperm released or deposited into water by aquatic animals. As a verb, *to spawn* refers to the process of releasing the eggs and sperm, and the act of both sexes is called spawning. Most aquatic animals, except for aquatic mammals and reptiles, reproduce through the process of spawning.

Spawn consists of the reproductive cells (gametes) of many aquatic animals, some of which will become fertilized and produce offspring. The process of spawning typically involves females releasing ova (unfertilized eggs) into the water, often in large quantities, while males simultaneously or sequentially release spermatozoa (milt) to fertilize the eggs.

Most fish reproduce by spawning, as do most other aquatic animals, including crustaceans such as crabs and shrimps, molluscs such as oysters and squid, echinoderms such as sea urchins and sea cucumbers, amphibians such as frogs and newts, aquatic insects such as mayflies and mosquitoes and corals, which are actually small aquatic animals—not plants. Fungi, such as mushrooms, are also said to "spawn" a white, fibrous matter that forms the matrix from which they grow.

There are many variations in the way spawning occurs, depending on sexual differences in anatomy, how the sexes relate to each other, where and how the spawn is released and whether or how the spawn is subsequently guarded.

Overview

Pacific salmon are semelparous or "big bang" spawners, which means they die shortly after spawning

The pickled and dehydrated roe of mullet

Marine animals, and particularly bony fish, commonly reproduce by *broadcast spawning*. This is an external method of reproduction where the female releases many unfertilised eggs into the water. At the same time, a male or many males release a lot of sperm into the water which fertilises some of these eggs. The eggs contain a drop of nutrient oil to sustain the embryo as it develops inside the egg case. The oil also provides buoyancy, so the eggs float and drift with the current. The strategy for survival of broadcast spawning is to disperse the fertilised eggs, preferably away from the coast into the relative safety of the open ocean. There the larvae develop as they consume their fat stores, and eventually hatch from the egg capsule into miniature versions of their parents. To survive, they must then become miniature predators themselves, feeding on plankton. Fish eventually encounter others of their own kind (conspecifics), where they form aggregations and learn to school.

Internally, the sexes of most marine animals can be determined by looking at the gonads. For example, male testes of spawning fish are smooth and white and account for up to 12% of the mass of the fish, while female ovaries are granular and orange or yellow, accounting for up to 70% of the fish's mass. Male lampreys, hagfish and salmon discharge their sperm into the body cavity where it is expelled through pores in the abdomen. Male sharks and rays can pass sperm along a duct into a seminal vesicle, where they store it for a while before it is expelled, while teleosts usually employ separate sperm ducts.

Externally, many marine animals, even when spawning, show little sexual dimorphism (difference in body shape or size) or little difference in colouration. Where species are dimorphic, such as sharks or guppies, the males often have penis-like intromittent organs in the form of a modified fin.

A species is semelparous if its individuals spawn only once in their lifetime, and iteroparous if its individuals spawn more than once. The term semelparity comes from the Latin *semel*, once, and *pario*, to beget, while iteroparity comes from *itero*, to repeat, and *pario*, to beget.

Semelparity is sometimes called "big bang" reproduction, since the single reproductive event of semelparous organisms is usually large and fatal to the spawners. The classic example of a semelparous animal is the Pacific salmon,which lives for many years in the ocean before swimming to the freshwater stream of its birth, spawning, and then dying. Other spawning animals which are semelparous include mayflies, squid, octopus, smelt, capelin and some amphibians. Semelparity is often associated with r-strategists. However, most fish and other spawning animals are iteroparous.

When the internal ovaries or egg masses of fish and certain marine animals are ripe for spawning they are called roe. Roe from certain species, such as shrimp, scallop, crab and sea urchins, are sought as human delicacies in many parts of the world. *Caviar* is a name for the processed, salted roe of non-fertilized sturgeon. The term *soft roe* or *white roe* denotes fish milt. Lobster roe is called *coral* because it turns bright red when cooked. Roe (reproductive organs) are usually eaten either raw or briefly cooked.

"The reproductive behaviour of fishes is remarkably diversified: they may be oviparous (lay eggs), ovoviviparous (retain the eggs in the body until they hatch), or viviparous (have a direct tissue connection with the developing embryos and give birth to live young). All cartilaginous fishes—the elasmobranches (e.g., sharks, rays, and skates)—employ internal fertilization and usually lay large, heavy-shelled eggs or give birth to live young. The most characteristic features of the more primitive bony fishes is the assemblage of polyandrous (many males) breeding aggregations in open water and the absence of parental care..."

There are two main reproduction methods in fish. The first method is by laying eggs and the second by *live-bearing* (producing their young alive).

*   In the first method, the female fish lays eggs either on the sea floor or on the leaves of an aquatic plant. A male fish fertilizes the eggs, and both then work together to protect the eggs/babies from danger until they can defend themselves.

*   In the second method, the male fish uses its anal fin to transmit sperm into the female fish and fertilize the fish eggs. Later, the female gives live birth to her fry.

Sexual Strategies

Basic Strategies

The four basic mating systems		
	Single female	**Multiple females**
Single male	Monogamy	Polygyny
Multiple males	Polyandry	Polygynandry

Monogamy occurs when one male mates with one female exclusively. This is also called *pair spawning*. Most fish are not monogamous, and when they are, they often alternate with non-monogamous behaviours. Monogamy can occur when feeding and breeding grounds are small, when it is difficult for fish to find partners, or when both sexes look after the young. Many tropical cichlids, which rear their young together in locations where they must fiercely defend against competitors and predators are monogamous. "In some pipefishes and seahorses, development of eggs takes a long time before the female can place them in the brood pouch of a male, where they are fertilized. While the male is pregnant, the female starts a new batch of eggs, which are ready at about the same time that the male gives birth to the young from the previous mating. This close timing of development promotes monogamy, especially if the likelihood of encountering another potential mate is low."

Cutthroat trout are monogamous pair spawners

The anglerfish *Haplophryne mollis* is polyandrous.
This female is trailing the atrophied remains of males she has encountered

Polygyny occurs when one male gets exclusive mating rights with multiple females. In polygyny a large conspicuous male usually defends females from other males or defends a breeding site. The females choose large males that are successfully defending prime breeding sites which the females find attractive. For example, sculpin males defend "caves" underneath rocks which are suitable for the incubation of embryos.

Another way males get to mate with several females is through the use of leks. Leks are places where many fish come together, and the males display to each another. Based on these displays, each female then selects the male they want to be their mate. For example, among the cichlid *Cyrtocara eucinosto-mus* in Lake Malawi, up to 50,000 large and colourful males display together on a lek four kilometres long. The females, which are mouth brooders, choose which male they want to fertilize their eggs.

Polyandry occurs when one female gets exclusive mating rights with multiple males. This is not common, but it does happen among fish like clownfish that change their sex. It can also happen when males do the brooding but can cannot handle all the eggs the female produce, such as with some pipefish.

The males in some deep sea anglerfishes are much smaller than the females. When they find a female they bite into her skin, releasing an enzyme that digests the skin of their mouth and her body and fusing the pair down to the blood-vessel level. The male then slowly atrophies, losing first his digestive organs, then his brain, heart, and eyes, ending as nothing more than a pair of gonads, which release sperm in response to hormones in the female's bloodstream indicating egg release. This extreme sexual dimorphism ensures that, when the female is ready to spawn, she has a mate immediately available. A single anglerfish female can "mate" with many males in this manner.

Polygynandry occurs when multiple males mate indiscriminately with multiple females. This mutual promiscuity is the approach most commonly used by spawning animals, and is perhaps the "original fish mating system." Common examples are forage fish, such as herrings, which form huge mating shoals in shallow water. The water becomes milky with sperm and the bottom is draped with millions of fertilized eggs.

Cuckoldry

Small male bluegill sunfishes cuckold large males by adopting *sneaker* or *satellite* strategies

Alternate male strategies which allow small males to engage in cuckoldry can develop in species where spawning is dominated by large and aggressive males. Cuckoldry is a variant of polyandry, and can occur with *sneak spawners* (sometimes called *streak spawners*). A sneak spawner is a male that rushes in to join the spawning rush of a spawning pair. A spawning rush occurs when a fish makes a burst of speed, usually on a near vertical incline, releasing gametes at the apex, followed by a rapid return to the lake or sea floor or fish aggregation. Sneaking males do not take part in courtship. In salmon and trout, for example, *jack males* are common. These are small silvery males that migrate upstream along with the standard, large, hook-nosed males and that spawn by sneaking into redds to release sperm simultaneously with a mated pair. This behaviour is an evolutionarily stable strategy for reproduction, because it is favoured by natural selection just like the "standard" strategy of large males.

Cuckoldry occurs in many fish species, including dragonets, parrotfishes and wrasses on tropical reefs and the bluegill sunfish in fresh water. Sneaker males that become too large to hide effectively become *satellite males*. With bluegill sunfish, satellite males mimic the behaviour and colouration of the females. They hover over a nest containing a pair of courting sunfish, and gradually descend to reach the pair just as they spawn. Males may need to be 6 or 7 years old to function capably as parental males, but may be able to function as sneaker or satellite males when they are as young as 2 or 3 years old. The smaller satellite and sneaker males may get mauled by the more powerful parental males, but they spawn when they are younger and they do not put energy into parental care.

Hermaphroditism

Hermaphroditism occurs when a given individual in a species possesses both male and female reproductive organs, or can alternate between possessing first one, and then the other. Hermaphroditism is common in invertebrates but rare in vertebrates. It can be contrasted with gonochorism, where each individual in a species is either male or female, and remains that way throughout their lives. Most fish are gonochorists, but hermaphroditism is known to occur in 14 families of teleost fishes.

Usually hermaphrodites are *sequential*, meaning they can switch sex, usually from female to male (protogyny). This can happen if a dominant male is removed from a group of females. The largest female in the harem can switch sex over a few days and replace the dominant male. This is found amongst coral reef fishes such as groupers, parrotfishes and wrasses. It is less common for a male to switch to a female (protandry). As an example, most wrasses are protogynous hermaphrodites within a haremic mating system. Hermaphroditism allows for complex mating systems. Wrasses exhibit three different mating systems: polygynous, lek-like, and promiscuous mating systems. Group spawning and pair spawning occur within mating systems. The type of spawning that occurs depends on male body size. Labroids typically exhibit broadcast spawning, releasing high amounts of planktonic eggs, which are broadcast by tidal currents; adult wrasses have no interaction with offspring. Wrasse of a particular subgroup of the Labridae family Labrini do not exhibit broadcast spawning.

Less commonly hermaphrodites can be *synchronous*, meaning they simultaneously possess both ovaries and testicles and can function as either sex at any one time. Black hamlets "take turns releasing sperm and eggs during spawning. Because such egg trading is advantageous to both individuals, hamlets are typically monogamous for short periods of time–an unusual situation in fishes." The sex of many fishes is not fixed, but can change with physical and social changes to the environment where the fish lives.

Particularly among fishes, hermaphroditism can pay off in situations where one sex is more likely to survive and reproduce, perhaps because it is larger. Anemone fishes are sequential hermaphrodites which are born as males, and become females only when they are mature. Anemone fishes live together monogamously in an anemone, protected by the anemone stings. The males do not have to compete with other males, and female anemone fish are typically larger. When a female dies a juvenile (male) anemone fish moves in, and "the resident male then turns into a female and reproductive advantages of the large female–small male combination continue". In other fishes sex changes are reversible. For example, if some gobies are grouped by sex (male or female), some will switch sex.

Unisexuality

Unisexuality occurs when a species is all-male or all-female. Unisexuality occurs in some fish species, and can take complex forms. *Squalius alburnoides*, a minnow found in several river basins in Portugal and Spain, appears to be an all-male species. The existence of this species illustrates the potential complexity of mating systems in fish. The species originated as a hybrid between two species, and is diploid, but not hermaphroditic. It can have triploid and tetraploid forms, including all-female forms that reproduce mainly through hybridogenesis.

It is rare to find true parthenogenesis in fishes, where females produce female offspring with no input from males. All-female species include the Texas silverside, *Menidia clarkhubbsi* as well as a complex of Mexican mollies. Parthenogenesis has been recently observed in hammerhead sharks and blacktip sharks. It is also known to occur in crayfish and amphibians.

Spawning Strategies

This section is patterned after a classification of the spawning behaviours of fish by Balon (1975, 1984) into reproductive guilds. This classification is based on how the eggs are fertilized (internal or external spawners), where the eggs are deposited (pelagic or benthic spawners), and whether and how the parents look after the eggs after spawning (bearers, guarders and nonguarders).

Nonguarders

Nonguarders do not protect their eggs and offspring after spawning

Open Substrate Spawners

Nonguarders: Open Substrate Spawners

- Pelagic spawners
- Benthic spawners
  - o Spawners on coarse bottoms
    - Pelagic free embryo and larvae
    - Benthic free embryo and larvae
  - o Spawners on plants
    - Obligatory
    - Nonobligatory
  - o Spawners on fine substrates
- Terrestrial spawners

Pike usually spawn on vegetation flooded by high water. Their eggs adhere to the stems of plants.

Nonguarders: Brood Hiders

- Benthic spawners

- Crevice spawners

- Spawners on invertebrates

- Beach spawners

Bitterlings transfer responsibility for the care of their young to mussels.
This male bitterling is exhibiting spawning colours

Open substrate spawners scatter their eggs in the environment. They usually spawn in shoals without complex courtship rituals, and males outnumber females.

Broadcast spawners: release their gametes (sperm and eggs) into open water for external fertilisation. There is no subsequent parental care. About 75% of coral species are broadcasters, the majority of which are hermatypic, or reef-building corals.

- Pelagic spawners: a type of broadcast spawners, spawn in the open sea, mostly near the surface. They are usually pelagic fish such as tuna and sardines. Some demersal fish leave the bottom to spawn pelagically, particularly coral reef fish such as parrotfish and wrasses. Pelagic spawning means water currents widely disperse the young. The eggs, embryos and larvae of pelagic spawners contain oil globules or have a high water content. As a result, they are buoyant and are widely dispersed by currents. The downside is that mortality is high, because they can be eaten so easily by pelagic predators or they can drift into unsuitable areas. Females compensate by spawning large numbers of eggs and extending their spawning periods. Pelagic spawners that live in or around coral reefs can spawn a small number of eggs almost daily over a period of months. These fishes have complex breeding behaviours including sex changes, harems, leks and territoriality.

- Benthic spawners: deposit their spawn on or near the bottom of the sea (or lake). They are usually demersal fish such as cod and flatfish. These species typically spawn without ceremony; they do not engage in elaborate courtship rituals. Each female is usually followed by several males who fertilize the eggs as they are released. Various strategies ensure the eggs and embryos remain in place, and do not drift with the current. The eggs can adhere to other eggs or to whatever they are deposited on, or the eggs can be laid in long strings which are wrapped around plants or rocks. Some eggs take on water after they are released, so they can be dropped into cracks where they swell and wedge themselves in place.

o   Egg scatterers: scatter adhesive or non-adhesive eggs to fall to the substrate, into plants, or float to the surface. These species do not look after their brood and even eat their own eggs. These are often schooling fish which spawn in groups or pairs, often laying a large number of small eggs. The fry hatch quickly.

o   Egg depositers: deposit eggs on a substrate (tank glass, wood, rocks, plants). Egg depositors usually lay fewer eggs than egg-scatterers, although the eggs are larger. Egg depositors fall into two groups: those that care for their eggs, and those that do not. Among egg depositors that care for their eggs are cichlids and some catfish. Egg depositors that care for their young can be divided into two groups: cavity spawners and open spawners.

o   Cavity spawners: lay eggs in a cave or cavity. These fish form pairs and have advanced brood care where the eggs are defended and cleaned. The eggs take a few days to hatch, and the fry are often guarded by the parents. Various catfish, Cyprinidae, and killifish make up the majority. Cavity spawners can be contrasted with open (shelter) spawners, which lay their eggs on an open surface.

Brood Hiders

Brood hiders hide their eggs but do not give parental care after they have hidden them. Brood hiders are mostly benthic spawners that bury the fertilized eggs. For example, among salmon and trout the female digs a nest with her tail in gravel. These nests are called redds. The female then lays her eggs while the male fertilizes them, while both fish defend the redd if necessary from other members of the same species. Then the female buries the nest, and the nest site is abandoned. In North America, some minnows build nests out of piles of stones rather than dig holes. The minnow males have tubercles on their head and body which they use to help them defend the nest site.

•   Egg buriers - can inhabit waters that dry up at some time of the year. An example are annual killifish which lay their eggs in mud. The parents mature quickly and lay their eggs before dying when the water dries up. The eggs remain in a dormant stage until rains stimulate hatching.

Bitterlings have a remarkable reproduction strategy where parents transfer responsibility for the care of their young to mussels. The female extends her ovipositor into the mantle cavity of the mussel and deposits her eggs between the gill filaments. The male then ejects his sperm into the mussel's inhalant water current and fertilization takes place within the gills of the host. The same female may use a number of mussels, and she deposits only one or two yellow, oval eggs into each. Early developmental stages are protected from predation within the body of the mussel. After 3 to 4 weeks larvae swim away from the host to continue life on their own.

Guarders

Guarders: Substrate Spawners

- Rock tenders

- Plant tenders

- Terrestrial tenders

- Pelagic tenders

Damselfish are substrate spawners. This one keeps her spawn in a gastropod shell

Guarders: Nest Spawners

- Rock and gravel nesters

- Sand nesters

- Plant-material nesters

  o Gluemakers

  o Nongluemakers

- Bubble nesters

- Hole nesters

- Misc-materials nesters

- Anemone nesters

Guarders protect their eggs and offspring after spawning by practicing *parental care* (also called *brood care*). Parental care is an "investment by parents in offspring that increases the offsprings' chances of surviving (and hence reproducing). In fish, parental care can take a variety of forms including guarding, nest building, fanning, splashing, removal of dead eggs, retrieval of straying fry, external egg carrying, egg burying, moving eggs or young, ectodermal feeding, oral brooding, internal gestation, brood-pouch egg carrying, etc."

The stickleback glues plant material to make its nest

Territorial behaviour is generally necessary for guarders, and the embryos are almost always guarded by males (apart from cichlids). There is a need to be territorial because looking after embryos usually includes defending the site where they are being looked after. It also often means there is competition for the best egg-laying sites. Elaborate courtship behaviour is usual among guarders.

Guarding males keep the embryos safe from predators, keep oxygen levels high by fanning water currents, and keep the area free from dead embryos and debris. They protect the embryos until they hatch, and often look after the larval stages as well. The time spent guarding can range from a few days to several months.

Substrate Spawners

Some guarders build nests (*nest spawners*) and some do not (*substrate spawners*), though the difference between the two groups can be small. Substrate spawners clean off a suitable area of surface suitable for egg laying, and look after the area, but they do not actively build a nest.

Baby paradise fish just hatched, gathered under the surface of a bubble nest

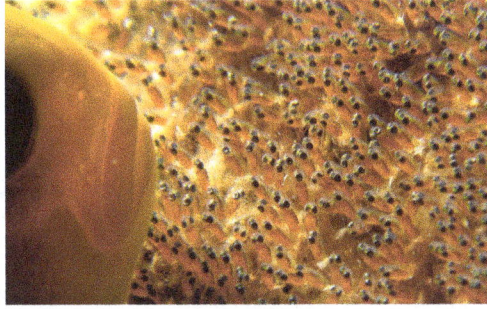

Anemone fish nest in an anemone. Here a male is protecting spawn produced by his partner.

Bearers

Bearers: External

- Transfer brooders

- Auxiliary brooders

- Mouth brooders

- Gill-chamber brooders

- Pouch brooders

Male seahorsees are pouch brooders

A female cichlid mouthbrooding fry which can be seen looking out her mouth

Bearers are fish that carry their embryos (and sometimes their young) around with them, either externally or internally.

External Bearers

Mouth brooders - carry eggs or larvae in their mouth. Mouth brooders can be ovophiles or larvo-philes. Ovophile or egg-loving mouth-brooders lay their eggs in a pit, which are sucked up into the mouth of the female. The small number of large eggs hatch in the mother's mouth, and the fry remain there for a period of time. Fertilization often occurs with the help of egg-spots, which are colorful spots on the anal fin of the male. When the female sees these spots, she tries to pick up the egg-spots, but instead gets a mouthful of sperm, fertilizing the eggs in her mouth. Many cichlids and some labyrinth fish are ovophile mouthbrooders. Larvophile or larvae-loving mouth-brooders lay their eggs on a substrate and guard them until the eggs hatch. After hatching, the female picks up the fry and keeps them in her mouth. When the fry can fend for themselves, they are released. Some eartheaters are larvophile mouthbrooders.

Internal Bearers

Facultative Internal Bearers

The beginning of the evolutionary process of livebearing starts with *facultative* (optional) internal bearing. The process occurs in several species of oviparous (egg-laying) killifishes which spawn in the normal way on the substrate, but in the process accidentally fertilize eggs which the female retains and does not spawn. These eggs are spawned later, usually without allowing much time for embryonic development.

Bearers: Internal

- Facultative internal bearers

- Obligate internal bearers

- Livebearers

Guppies are livebearers. This one has been pregnant for about 26 days.

Obligate Internal Bearers

The next step in the evolution of livebearing is *obligate* (by necessity) internal bearing, where the female retains all the embryos. "The only source of nutrition for these embryos, however, is the egg yolk, as in externally spawned eggs. This situation, also referred to as ovoviviparity, is characteristic of marine rock fishes and the Lake Baikal sculpins. This strategy allows these fish to have fecundities approaching those of pelagic fish with external fertilization, but it also enables them to protect the young during their most vulnerable stage of development. By contrast, sharks and rays using this strategy produce a relatively small number of embryos and retain them for a few weeks to 16 months or longer. The shorter times spans are characteristic of species that eventually deposit their embryos in the environment, surrounded by a horny capsule; whereas the longer periods are characteristic of sharks that retain the embryos until they are ready to emerge as actively swimming young."

Viviparous Fish

However, some fish do not fit these categories. The livebearing largespring gambusia (*Gambusia geiseri*) was thought to be ovoviviparous until it was shown in 2001 that the embryos received nutrients from the mother.

Spawning Grounds

Capelin migrate huge distances to their spawning grounds

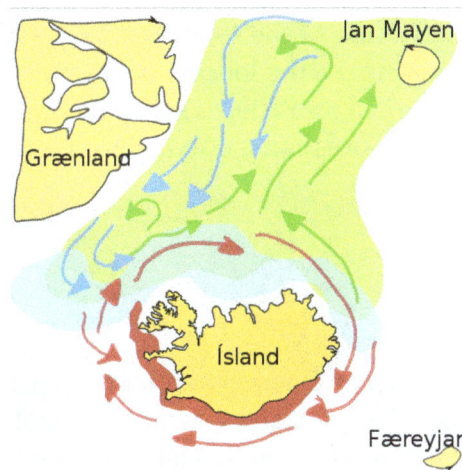

Migration of capelin around Iceland. Capelin on the way to feeding grounds is green,
capelin on the way back is blue, the breeding grounds are red.

Spawning grounds are the areas of water where aquatic animals spawn, or produce their eggs. After spawning, the spawn may or may not drift to new grounds which become their nursery grounds. Many species undertake migrations each year, and sometimes great migrations, to reach their spawning grounds. For example, lakes and river watersheds can be major spawning grounds for anadromous fish such as salmon. These days, it is often necessary to construct fish ladders and other bypass systems so salmon can navigate their way past hydroelectric dams or other obstructions such as weirs on their way to spawning grounds. Coastal fish often use mangroves and estuaries as spawning grounds, while reef fish can find adjacent seagrass meadows that make good spawning grounds. Short-finned eels can travel anything up to three or four thousand kilometres to their spawning ground in deep water somewhere in the Coral Sea.

Forage fish often make great migrations between their spawning, feeding and nursery grounds. Schools of a particular stock usually travel in a triangle between these grounds. For example, one stock of herrings have their spawning ground in southern Norway, their feeding ground in Iceland, and their nursery ground in northern Norway. Wide triangular journeys such as these may be important because forage fish, when feeding, cannot distinguish their own offspring.

Capelin are a forage fish of the smelt family found in the Atlantic and Arctic oceans. In summer, they graze on dense swarms of plankton at the edge of the ice shelf. Larger capelin also eat krill and other crustaceans. The capelin move inshore in large schools to spawn and migrate in spring and summer to feed in plankton rich areas between Iceland, Greenland, and Jan Mayen. The migration is affected by ocean currents. Around Iceland maturing capelin make large northward feeding migrations in spring and summer. The return migration takes place in September to November. The spawning migration starts north of Iceland in December or January.

The diagram on the right shows the main spawning grounds and larval drift routes. Capelin on the way to feeding grounds is coloured green, capelin on the way back is blue, and the breeding grounds are red. In a paper published in 2009, researchers from Iceland recount their application of an interacting particle model to the capelin stock around Iceland, successfully predicting the spawning migration route for 2008.

Referred to as "the greatest shoal on earth", the sardine run occurs when millions of sardines migrate from their spawning grounds south of the southern tip of Africa northward along the Eastern Cape coastline. Chinook salmon make the longest freshwater migration of any salmon, over 3,000 kilometres (1,900 mi) up the Yukon River to spawning grounds upstream of Whitehorse, Yukon. Some green sea turtles swim more than 2,600 kilometres (1,600 mi) to reach their spawning grounds.

Examples

Goldfish

Goldfish, like all cyprinids, are egg-layers. They usually start breeding after a significant temperature change, often in spring. Males chase females, prompting them to release their eggs by bumping and nudging them. As the female goldfish spawns her eggs, the male goldfish stays close behind fertilizing them. Their eggs are adhesive and attach to aquatic vegetation. The eggs hatch within 48 to 72 hours. Within a week or so, the fry begins to assume its final shape, although a year may pass before they develop a mature goldfish colour; until then they are a metallic brown like their wild ancestors. In their first weeks of life, the fry grow quickly—an adaptation born of the high risk of getting devoured by the adult goldfish.

Carp

A member of the Cyprinidae family, carp spawn in times between April and August, largely dependent upon the climate and conditions they live in. Oxygen levels of the water, availability of food, size of each fish, age, number of times the fish has spawned before and water temperature are all factors known to effect when and how many eggs each carp will spawn at any one time.

Siamese Fighting Fish

Prior to spawning, male Siamese fighting fish build bubble nests of varying sizes at the surface of the water. When a male becomes interested in a female, he will flare his gills, twist his body, and spread his fins. The female darkens in colour and curves her body back and forth. The act of spawning takes place in a "nuptial embrace" where the male wraps his body around the female, each embrace resulting in the release of 10-40 eggs until the female is exhausted of eggs. The male, from his side, releases milt into the water and fertilization takes place externally. During and after spawning, the male uses his mouth to retrieve sinking eggs and deposit them in the bubble nest (during mating the female sometimes assists her partner, but more often she will simply devour all the eggs that she manages to catch). Once the female has released all of her eggs, she is chased away from the male's territory, as it is likely that she'll eat the eggs due to hunger. The eggs then remain in the male's care. He keeps them in the bubble nest, making sure none fall to the bottom and repairing the nest as needed. Incubation lasts for 24–36 hours, and the newly hatched larvae remain in the nest for the next 2–3 days, until their yolk sacs are fully absorbed. Afterwards the fry leave the nest and the free-swimming stage begins.

One-day-old Siamese fighting fish larvae in a bubble nest - their yolk sacs have not yet been absorbed

Crustaceans

California spiny lobster

Copepods

Copepods are tiny crustaceans which usually reproduce either by *broadcast spawning* or by *sac spawning*. Broadcasting copepods scatter their eggs into the water, but sac spawners lay their eggs into an ovigerous sac. Sac spawners spawn few but relatively large eggs that develop slowly. By contrast, broadcast spawners spawn numerous small eggs that develop rapidly. However, the shorter hatch times that result from broadcasting are not short enough to compensate for the higher mortality compared to sac spawners. To produce a given number of hatched eggs, broadcasters must spawn more eggs than sac spawners.

Spiny Lobsters

After mating, the fertilized eggs of the California spiny lobster are carried on the female's pleopods until they hatch, with between 120,000 and 680,000 carried by a single female. The eggs begin coral red, but darken as they develop to a deep maroon. When she is carrying the eggs, the female is said to be "berried". The eggs are ready to hatch after 10 weeks, and spawning takes place from May to August. The larvae that hatch (called *phyllosoma* larvae) do not resemble the adults. Instead, they are flat, transparent animals around 14 mm (0.55 in) long, but as thin as a sheet of paper. The larvae feed on plankton, and grow through ten molts into ten further larval stages, the last of which is around 30–32 mm (1.2–1.3 in) long. The full series of larval molts takes around 7 months, and when the last stage molts, it metamorphoses into the *puerulus* state, which is a juvenile form of the adult, though still transparent. The puerulus larvae settle to the sea floor when the water is near its maximum temperature, which in Baja California is in the fall.

Egg-bearing female lobsters migrate inshore from deeper waters to hatch their eggs, though they do not have specific spawning grounds. These lobster migrations can occur in close single-file formation "lobster trains".

Molluscs

Pacific oyster

Pacific Oysters

Oysters are *broadcast spawners*, that is, eggs and sperm are released into open water where fertilisation occurs. They are protandric; during their first year they spawn as males by releasing sperm into the water. As they grow over the next two or three years and develop greater energy reserves, they spawn as females by releasing eggs. Bay oysters usually spawn by the end of June. An increase in water temperature prompts a few oysters to spawn. This triggers spawning in the rest, clouding the water with millions of eggs and sperm. A single female oyster can produce up to 100 million eggs annually. The eggs become fertilized in the water and develop rapidly into planktonic larvae. which eventually find suitable sites, such as another oyster's shell, on which to settle. Attached oyster larvae are called spat. Spat are oysters less than 25 millimetres (0.98 in) long.

The Pacific oyster usually has separate sexes. Their sex can be determined by examining the gonads, and it can change from year to year, normally during the winter months. In certain environmental conditions, one sex is favoured over the other. Protandry is favoured in areas of high food abundance and protogyny occurs in areas of low food abundance. In habitats with a high food supply, the sex ratio in the adult population tends to favour females, and areas with low food abundances tend to have a larger proportion of male adults. Spawning in the Pacific oyster occurs at 20 °C (68 °F). This species is very fecund, with females releasing about 50–200 million eggs in regular intervals (at a rate of 5–10 times a minute) in a single spawning. Once released from the gonads, the eggs move through the suprabranchial chambers (gills), are then pushed through the gill ostia into the mantle chamber, and are finally released in the water, forming a small cloud. In males, the sperm is released at the opposite end of the oyster, along with the normal exhalent stream of water. A rise in water temperature is thought to be the main cue in the initiation of spawning, as the onset of higher water temperatures in the summer results in earlier spawning in the Pacific oyster.

The larvae of the Pacific oyster are planktotrophic, and are about 70 μm at the prodissoconch 1 stage. The larvae move through the water column via the use of a larval foot to find suitable settlement locations. They can spend several weeks at this phase, which is dependent on water temperature, salinity and food supply. Over these weeks, larvae can disperse great distances by water currents before they metamorphose and settle as small spat. Similar to other oyster species, once the Pacific oyster larvae find a suitable habitat, they attach to it permanently using cement secreted from a gland in their foot. After settlement, the larvae metamorphose into juvenile spat. The growth rate is very rapid in optimum environmental conditions, and market size can be achieved in 18 to 30 months.

Egg cases laid by a female squid

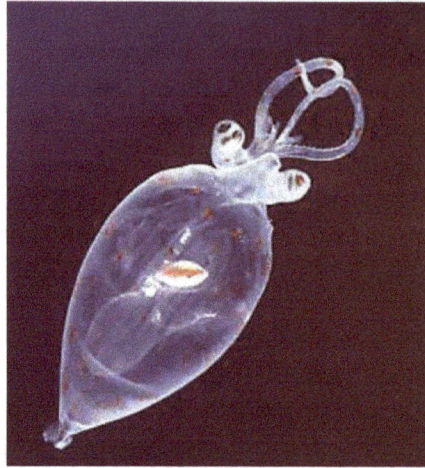
A juvenile squid

Cephalopods

Cephalopods, such as squid and octopuses, have prominent heads and a set of arms (tentacles) modified from the primitive foot of molluscs. All cephalopods are sexually dimorphic. However, they lack external sexual characteristics, so they use colour communication. A courting male approaches a likely looking mate flashing his brightest colours, often in rippling displays. If the other cephalopod is female and receptive, her skin will change colour to become pale, and mating will occur. If the other cephalopod remains brightly coloured, it is taken as a warning.

All cephalopods reproduce by spawning eggs. Most cephalopods use semi-internal fertilization where the male places his gametes inside the female's mantle cavity to fertilize the ova in the female's single ovary. The "penis" in most male cephalopods is a long and muscular end of the gonoduct used to transfer spermatophores to a modified sperm-carrying arm called a hectocotylus. That in turn is used to transfer the spermatophores to the female. In species where the hectocotylus is missing, the "penis" is long and able to extend beyond the mantle cavity and transfers the spermatophores directly to the female. In many cephalopods, mating occurs head to head and the male may simply transfer sperm to the female. Others may detach the sperm-carrying arm and leave it attached to the female. Deep water squid have the greatest known penis length relative to body size of all mobile animals, second in the entire animal kingdom only to certain sessile barnacles. Penis elongation in the greater hooked squid may result in a penis that is as long as the mantle, head and arms combined.

A greater hooked squid with an erect penis 67 cm long

Some species brood their fertilized eggs: female paper nautilus construct shelters for the young, while Gonatiid squid carry a larva-laden membrane from the hooks on their arms. Other cephalopods deposit their young under rocks and aerate them with their tentacles hatching. Mostly the eggs are left to their own devices; many squid lay sausage-like bunches of eggs in crevices or occasionally on the sea floor. Cuttlefish lay eggs separately in cases and attach them to coral or algal fronds. Like Pacific salmon, cephalopods are mostly semelparous, spawning many small eggs in one batch and then dying. Cephalopods usually live fast and die young. Most of the energy extracted from their food is used for growing, and they mature rapidly to their adult size. Some gain as much as 12% of their body mass each day. Most live for one to two years, reproducing and then dying shortly thereafter.

Echinoderms

Sea urchins have five gonads. These gonads (roe) are a sought after as a delicacy.

Echinoderms are marine animals, widespread in all oceans, but not found in fresh water. Just below their skin is an endoskeleton composed of calcareous plates or ossicles.

Sea Urchins

Sea urchins are spiky echinoderms with spherical bodies which usually contain five gonads. They move slowly, feed mostly on seaweed, and are important for the diet of sea otters. Sea urchins are dioecious, having separate male and female sexes, although there is generally no easy way to distinguish the two. The gonads are lined with muscles underneath the peritoneum, and these allow the animal to squeeze its gametes through the duct and into the surrounding sea water, where fertilization takes place. Their roe (male and female gonads) is soft and melting, with a colour ranging from orange to pale yellow, and is sought after as a human delicacy in many parts of the world.

Sea Cucumbers

Sea cucumbers are leathery echinoderms with elongated bodies which contain a single, branched gonad. They are found on the sea floor worldwide, and occur in great numbers on the deep sea floor where they often make up the majority of the animal biomass. They feed on plankton and decaying organic debris found at the sea bottom, catching food that flows by with their open tentacles or sifting through bottom sediments. Like sea urchins, most sea cucumbers reproduce by releasing sperm and ova into the ocean water. Depending on conditions, one organism can produce thousands of gametes.

Sea cucumbers have one gonad

Sea cucumbers are typically dioecious, with separate male and female individuals. The reproductive system consists of a single gonad, consisting of a cluster of tubules emptying into a single duct that opens on the upper surface of the animal, close to the tentacles. Many species fertilise their eggs internally. The fertilised egg develops in a pouch on the adult's body and eventually hatches as a juvenile sea cucumber. A few species brood their young inside the body cavity, giving birth through a small rupture in the body wall close to the anus. The remaining species develop their eggs into a free-swimming larva, usually after about three days of development. This larva swims by means of a long band of cilia wrapped around its body. As the larva grows it transforms into a barrel-shaped body with three to five separate rings of cilia. The tentacles are usually the first adult features to appear, before the regular tube feet.

Amphibious Animals

Amphibians have successfully solved most of the problems associated with exposure to air. But their reproductive system was and is linked to water, and it remains very fishlike. Almost all amphibians spawn in water and lay a great number of small eggs that hatch quickly into swimming larvae. The eggs do not need any complex protection against drying, because if the environment dries, the larvae are doomed as well as the eggs. Thus selection has acted to encourage the selection of suitable sites for laying eggs, rather than suitable devices for protecting eggs. Both fishes and amphibians may migrate long distances for spawning, and favoured sites are often disputed vigorously.

Common frogs sorting out their spawn

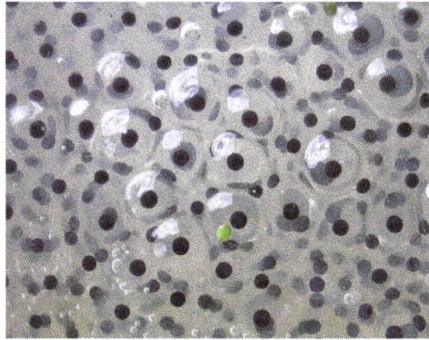
Frog spawn up close

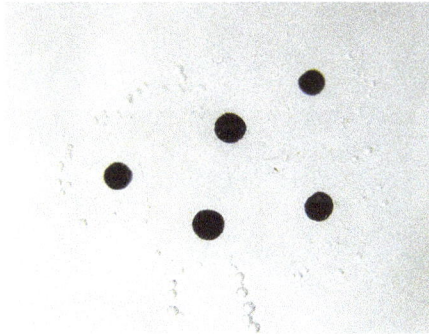
Frog spawn development

Amphibians are found in and around fresh water lakes and ponds, but not in marine environments. Examples are frogs and toads, salamanders, newts and caecilians (which resemble snakes). They are cold-blooded animals that metamorphose from a juvenile water-breathing form, usually to an adult air-breathing form, though mudpuppies retain juvenile gills in adulthood.

Frogs and Toads

Female frogs and toads usually spawn gelatinous egg masses containing thousands of eggs in water. Different species lay eggs in distinctive and identifiable ways. For example, the American toad lays long strings of eggs. The eggs are highly vulnerable to predation, so frogs have evolved many techniques to ensure the survival of the next generation. In colder areas the embryo is black to absorb more heat from the sun, which speeds up development. Most commonly, this involves synchronous reproduction. Many individuals will breed at the same time, overwhelming the actions of predators; the majority of the offspring will still die due to predation, but there is a greater chance some will survive. Another way in which some species avoid predators and the pathogens eggs are exposed to in ponds is to lay eggs on leaves above the pond, with a gelatinous coating designed to retain moisture. In these species the tadpoles drop into the water upon hatching. The eggs of some species laid out of water can detect vibrations of nearby predatory wasps or snakes, and will hatch early to avoid being eaten.

While the length of the egg stage depends on the species and environmental conditions, aquatic eggs generally hatch within one week. Unlike salamanders and newts, frogs and toads never become sexually mature while still in their larval stage. The hatched eggs continue life as tadpoles, which typically have oval bodies and long, vertically flattened tails. As a general rule, free living larvae are fully aquatic. They lack eyelids and have a cartilaginous skeleton, a lateral line system, gills for respiration (external gills at first, internal gills later) and tails with dorsal and ventral folds of skin for swimming. They quickly develop a gill pouch that covers the gills and the front legs; the lungs are also developed at an early stage as an accessory breathing organ. Some species which go through the metamorphosis inside the egg and hatch to small frogs never develop gills; instead there are specialised areas of skin that take care of respiration. Tadpoles also lack true teeth, but the jaws in most species usually have two elongate, parallel rows of small keratinized structures called keradonts in the upper jaw while the lower jaw has three rows of keradonts, surrounded by a horny beak, but the number of rows can be lower (sometimes zero), or much higher. Tadpoles feed on algae, including diatoms filtered from the water through the gills. Some species are carnivorous at the tadpole stage, eating insects, smaller tadpoles, and fish. Cannibalism has been observed among tadpoles. Early developers who gain legs may be eaten by the others, so the late bloomers survive longer.

Sea Turtles

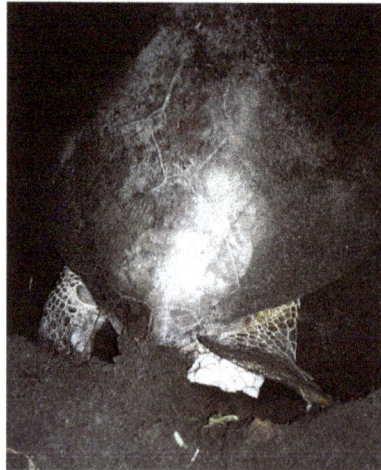
Sea turtle laying eggs

Sea turtles are amphibious reptiles, but they are not amphibians. Reptiles belong to the class Reptilia while amphibians belong to the class Amphibia. These are two distinct taxonomic groups. Reptiles have scales and leathery skins, while the skins of amphibians are smooth and porous. Unlike frogs, sea turtle eggs have tough, leathery shells which allow them to survive on land without drying out.

Some sea turtles migrate long distances between feeding and spawning grounds. Green turtles have feeding grounds along the Brazilian coast. Each year, thousands of these turtles migrate about 2,300 kilometres (1,400 mi) to their spawning ground, Ascension Island in the Atlantic, an island only 11 kilometres (6.8 mi) across. Each year the returning turtles dig between 6,000 and 15,000 nests, often returning to the same beach from where they hatched. Females usually mate every two to four years. Males on the other hand visit the breeding areas every year, attempting to mate. Green sea turtles' mating is similar to other marine turtles. Female turtles control the process. A few populations practice polyandry, although this does not seem to benefit hatchlings. After mating in the water, the female moves above the beach's high tide line where she digs a hole with her hind flippers and deposits her eggs. Litter size depends on the age of the female and species, but green turtle clutches range between 100 and 200. She then covers the nest with sand and returns to the sea.

Green turtle hatchling

At around 45 to 75 days, the eggs hatch during the night and the hatchlings instinctively head directly into the water. This is the most dangerous time in a turtle's life. As they walk, predators such as gulls and crabs grab them. A significant percentage never make it to the ocean. Little is known of the initial life history of newly hatched sea turtles. Juveniles spend three to five years in the open ocean before they settle as still-immature juveniles into their permanent shallow-water lifestyle. It is speculated that they take twenty to fifty years to reach sexual maturity. Individuals live up to eighty years in the wild. They are among the larger sea turtles, many more than a meter long and weighing up to 300 kilograms (660 lb).

Aquatic Insects

Aquatic insects also spawn. Mayflies "are famed for their short adult life. Some species have under an hour to mate and lay their eggs before they die. Their pre-adult stage, known as the subimago, may be even shorter - perhaps lasting just a few minutes before they moult into their adult form. Therefore a mayfly spends most of its life as a nymph, hidden from view under the water."

Corals

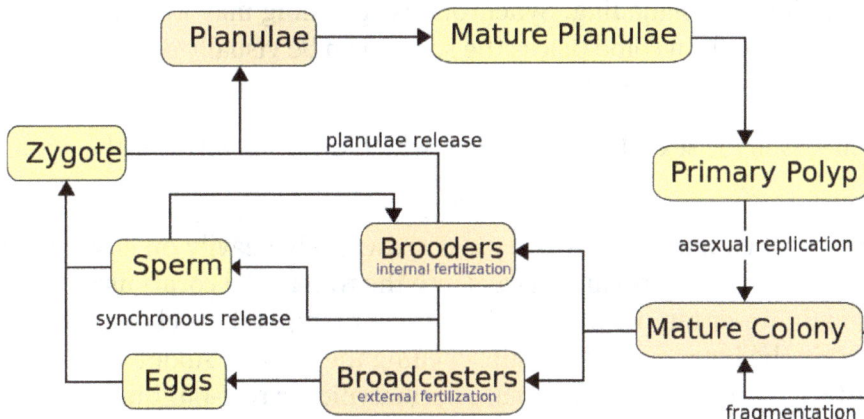

Life cycles of broadcasters and brooders

Corals can be both gonochoristic (unisexual) and hermaphroditic, each of which can reproduce sexually and asexually. Reproduction also allows corals to settle new areas.

Corals predominantly reproduce sexually. 25% of hermatypic corals (stony corals) form single sex (gonochoristic) colonies, while the rest are hermaphroditic. About 75% of all hermatypic corals "broadcast spawn" by releasing gametes—eggs and sperm—into the water to spread offspring. The gametes fuse during fertilization to form a microscopic larva called a planula, typically pink and elliptical in shape. A typical coral colony form several thousand larvae per year to overcome the odds against formation of a new colony.

Planulae exhibit positive *phototaxis*, swimming towards light to reach surface waters where they drift and grow before descending to seek a hard surface to which they can attach and establish a new colony. They also exhibit positive *sonotaxis*, moving towards sounds that emanate from the reef and away from open water. High failure rates afflict many stages of this process, and even though millions of gametes are released by each colony very few new colonies form. The time from spawning to settling is usually 2–3 days, but can be up to 2 months. The larva grows into a polyp and eventually becomes a coral head by asexual budding and growth.

A male star coral, *Montastraea cavernosa*, releases sperm into the water

Synchronous spawning is very typical on the coral reef and often, even when multiple species are present, all corals spawn on the same night. This synchrony is essential so that male and female gametes can meet. Corals must rely on environmental cues, varying from species to species, to determine the proper time to release gametes into the water. The cues involve lunar changes, sunset time, and possibly chemical signalling. Synchronous spawning may form hybrids and is perhaps involved in coral speciation. In some places the spawn can be visually dramatic, clouding the usually clear water with gametes, typically at night.

Corals use two methods for sexual reproduction, which differ in whether the female gametes are released:

- Broadcasters, the majority of which mass spawn, rely heavily on environmental cues, because they release both sperm and eggs into the water. The corals use long-term cues such as day length, water temperature, and/or rate of temperature change. The short-term cue is most often the lunar cycle, with sunset cuing the release. About 75% of coral species are broadcasters, the majority of which are hermatypic, or reef-building corals. The positively buoyant gametes float towards the surface where fertilization produces planula larvae. The larvae swim towards the surface light to enter into currents, where they usually remain for two days, but sometimes up to three weeks, and in one known case two months, after which they settle and metamorphose into polyps and form colonies.

- Brooders are most often ahermatypic (non-reef building) in areas of high current or wave action. Brooders release only sperm, which is negatively buoyant, and can harbor unfertilized eggs for weeks, lowering the need for mass synchronous spawning events, which do sometimes occur. After fertilization the corals release planula larvae which are ready to settle.

Fungi

Harvesting oyster mushroom *Pleurotus ostreatus* cultivated using spawns embedded in sawdust mixture placed in plastic containers

Fungi are not plants, and require different conditions for optimal growth. Plants develop through photosynthesis, a process that converts atmospheric carbon dioxide into carbohydrates, especially cellulose. While sunlight provides an energy source for plants, mushrooms derive all of their energy and growth materials from their growth medium, through biochemical decomposition processes. This does not mean that light is an unnecessary requirement, since some fungi use light as a signal to induce fruiting. However, all the materials for growth must already be present in the growth medium. Instead of seeds, mushrooms reproduce sexually during underground growth, and asexually through spores. Either of these can be contaminated with airborne microorganisms, which will interfere with mushroom growth and prevent a healthy crop. Mycelium, or actively growing mushroom culture, is placed on growth substrate to seed or introduce mushrooms to grow on a substrate. This is also known as inoculation, spawning or adding spawn. Its main advantages are to reduce chances of contamination while giving mushrooms a firm beginning.

Ichthyoplankton

Ichthyoplankton (from Greek, *ikhthus*, "fish" and *planktos*, "drifter") are the eggs and larvae of fish. They are usually found in the sunlit zone of the water column, less than 200 metres deep, which is sometimes called the epipelagic or photic zone. Ichthyoplankton are planktonic, meaning they cannot swim effectively under their own power, but must drift with the ocean currents. Fish eggs cannot swim at all, and are unambiguously planktonic. Early stage larvae swim poorly, but later stage larvae swim better and cease to be planktonic as they grow into juveniles. Fish larvae are part of the zooplankton that eat smaller plankton, while fish eggs carry their own food supply. Both eggs and larvae are themselves eaten by larger animals.

Fish can produce high numbers of eggs which are often released into the open water column. Fish eggs typically have a diameter of about 1 millimetre (0.039 in). The newly hatched young of oviparous fish are called larvae. They are usually poorly formed, carry a large yolk sac (for nourishment) and are very different in appearance from juvenile and adult specimens. The larval period in oviparous fish is relatively short (usually only several weeks), and larvae rapidly grow and change appearance and structure (a process termed metamorphosis) to become juveniles. During this transition larvae must switch from their yolk sac to feeding on zooplankton prey, a process which depends on typically inadequate zooplankton density, starving many larvae.

Ichthyoplankton can be a useful indicator of the state and health of an aquatic ecosystem. For instance, most late stage larvae in ichthyoplankton have usually been preyed on, so ichthyoplankton tends to be dominated by eggs and early stage larvae. This means that when fish, such as anchovies and sardines, are spawning, ichthyoplankton samples can reflect their spawning output and provide an index of relative population size for the fish. Increases or decreases in the number of adult fish stocks can be detected more rapidly and sensitively by monitoring the ichthyoplankton associated with them, compared to monitoring the adults themselves. It is also usually easier and more cost effective to sample trends in egg and larva populations than to sample trends in adult fish populations.

History

Interest in plankton originated in Britain and Germany in the nineteenth century when researchers discovered there were microorganisms in the sea, and that they could trap them with fine-mesh nets. They started describing these microorganisms and testing different net configurations. Ichthyoplankton research started in 1864 when the Norwegian government commissioned the marine biologist G. O. Sars to investigate fisheries around the Norwegian coast. Sars found fish eggs, particularly cod eggs, drifting in the water. This established that fish eggs could be pelagic, living in the open water column like other plankton. Around the beginning of the twentieth century, research interest in ichthyoplankton became more general when it emerged that, if ichthyoplankton was sampled quantitatively, then the samples could indicate the relative size or abundance of spawning fish stocks.

Sampling Methods

Retrieving a Plankton Sample

Research vessels collect ichthyoplankton from the ocean using fine mesh nets. The vessels either tow the nets through the sea or pump sea water onboard and then pass it through the net.

- There are many types of plankton tows:
    - Neuston net tows are often made at or just below the surface using a nylon mesh net fitted to a rectangular frame

- The PairoVET tow, used for collecting fish eggs, drops a net about 70 metres into the sea from a stationary research vessel and then drags it back to the vessel.

- Ring net tows involve a nylon mesh net fitted to a circular frame. These have largely been replaced by bongo nets, which provide duplicate samples with their dual-net design.

- The bongo tow drags nets shaped like bongo drums from a moving vessel. The net is often lowered to about 200 metres and then allowed to rise to the surface as it is towed. In this way, a sample can be collected across the whole photic zone where most ichthyoplankton is found.

- MOCNESS tows and Tucker trawls utilize multiple nets that are mechanically opened and closed at discrete depths in order to provide insights into the vertical distribution of the plankton

- The manta trawl tows a net from a moving vessel along the surface of the water, collecting larvae, such as grunion, mahi-mahi, and flying fish which live at the surface.

After the tow the plankton is flushed with a hose to the cod end (bottom) of the net for collection. The sample is then placed in preservative fluid prior to being sorted and identified in a laboratory.

- Plankton pumps: Another method of collecting ichthyoplankton is to use a Continuous Underway Fish Egg Sampler (CUFES). Water from a depth of about three metres is pumped onto the vessel and filtered with a net. This method can be used while the vessel is underway.

Developmental Stages

Ichthyoplankton researchers generally use the terminology and development stages introduced in 1984 by Kendall and others. This consists of three main developmental stages and two transitional stages.

Survival

Recruitment of fish is regulated by larval fish survival. Survival is regulated by prey abundance, predation, and hydrology. Fish eggs and larvae are eaten by many marine organisms. For example, they may be fed upon by marine invertebrates, such as copepods, arrow worms, jellyfish, amphipods, marine snails and krill. Because they are so abundant, marine invertebrates inflict high overall mortality rates. Adult fish also prey on fish eggs and larvae. For example, haddock were observed satiating themselves with herring eggs back in 1922. Another study found cod in a herring spawning area with 20,000 herring eggs in their stomachs, and concluded that they could prey on half of the total egg production. Fish also cannibalise their own eggs. For example, separate studies found northern anchovy (*Engraulis mordax*) were responsible for 28% of the mortality in their own egg population, while Peruvian anchoveta were responsible for 10% and South African anchovy (*Engraulis encrasicolus*) 70%.

The most effective predators are about ten times as long as the larvae they prey on. This is true regardless of whether the predator is a crustacean, a jellyfish, or a fish.

Dispersal

The larvae of the yellow tang can drift more than 100 miles and reseed in a distant location.

Fish larvae develop first an ability to swim up and down the water column for short distances. Later they develop an ability to swim horizontally for much longer distances. These swimming developments affect their dispersal.

In 2010, a group of scientists reported that fish larvae can drift on ocean currents and reseed fish stocks at a distant location. This finding demonstrates, for the first time, what scientists have long suspected but have never proven, that fish populations can be connected to distant populations through the process of larval drift.

The fish they chose to investigate was the yellow tang, because when a larva of this fish find a suitable reef it stays in the general area for the rest of its life. Thus, it is only as drifting larvae that the fish can migrate significant distances from where they are born. The tropical yellow tang is much sought after by the aquarium trade. By the late 1990s, their stocks were collapsing, so in an attempt to save them nine marine protected areas (MPAs) were established off the coast of Hawaii. Now, through the process of larval drift, fish from the MPAs are establishing themselves in different locations, and the fishery is recovering. "We've clearly shown that fish larvae that were spawned inside marine reserves can drift with currents and replenish fished areas long distances away," said one of the authors, the marine biologist Mark Hixon. "This is a direct observation, not just a model, that successful marine reserves can sustain fisheries beyond their borders."

Pregnancy in Fish

Pregnancy has been traditionally defined as the period during which developing embryos are incubated in the body after egg-sperm union. Although the term often refers to placental mammals, it has also been used in the titles of many international, peer-reviewed, scientific articles on fish, e.g. Consistent with this definition, there are several modes of reproduction in fish, providing different amounts of parental care. In ovoviviparity, there is internal fertilization and the young are born live but there is no placental connection or significant trophic (feeding) interaction; the mother's body maintains gas exchange but the unborn young are nourished by egg yolk. There are two types of viviparity in fish. In histotrophic viviparity, the zygotes develop in the female's oviducts, but she provides no direct nutrition; the embryos survive by eating her eggs or their unborn siblings. In hemotrophic viviparity, the zygotes are retained within the female and are provided with nutrients by her, often through some form of placenta.

A pregnant fish

In seahorses and pipefish, it is the male that becomes pregnant.

Types of Reproduction and Pregnancy

Pregnancy has been traditionally defined as the period during which developing embryos are incubated in the body after egg-sperm union. Despite strong similarities between viviparity in mammals, researchers have historically been reluctant to use the term "pregnancy" for non-mammals because of the highly developed form of viviparity in eutherians. Recent research into physiological, morphological and genetic changes associated with fish reproduction provide evidence that incubation in some species is a highly specialized form of reproduction similar to other forms of viviparity. Although the term "pregnancy" often refers to eutherian animals, it has also been used in the titles of many international, peer-reviewed, scientific articles on fish, e.g. five modes of reproduction can be differentiated in fish based on relations between the zygote(s) and parents:

- Ovuliparity: Fertilization of eggs is external; zygotes develop externally.

- Oviparity: Fertilization of eggs is internal; zygotes develop externally as eggs with large vitellus.

- Ovoviviparity: Fertilization is internal; zygotes are retained in the female (or male) but without major trophic (feeding) interactions between zygote and parents (there may be minor interactions, such as maintenance of water and oxygen levels). The embryos depend upon their yolk for survival.

There are two types of viviparity among fish.

- Histotrophic ("tissue eating") viviparity: The zygotes develop in the female's oviducts, but she provides no direct nutrition. The embryos survive by eating her eggs or their unborn siblings.

- Hemotrophic ("blood eating") viviparity: The zygotes are retained within the female and are provided with nutrients by her, often through some form of placenta.

Ovoviviparous Fish

Examples of ovoviviparous fish are many of the squaliform sharks, which include sand sharks, mackerel sharks, nurse sharks, requiem sharks, dog sharks and hammerheads, among others, and the lobe finned coelacanth. Some species of rockfish (*Sebastes*) and sculpins (Comephoridae) produce rather weak larvae with no egg membrane and are also, by definition, ovoviviparous. Ovoviviparity occurs in most live-bearing bony fishes (Poeciliidae).

Viviparous Fish

Viviparous fish include the families Goodeidae, Anablepidae, Jenynsiidae, Poeciliidae, Embiotocidae and some sharks (some species of the requiem sharks, Carcharinidae and the hammerheads, Sphyrnidae, among others). The halfbeaks, Hemiramphidae, are found in both marine and freshwaters and those species that are marine produce eggs with extended filaments that attach to floating or stationary debris, while those that are found in freshwater are viviparous with internal fertilization. The Bythitidae are also viviparous although one species, *Dinematichthys ilucoeteoides*, is reported to be ovoviviparous.

Aquarists commonly refer to ovoviviparous and viviparous fish as "livebearers". Examples include guppies, mollies, moonfish, platys, four-eyed fish and swordtails. All of these varieties exhibit signs of their pregnancy before the live fry are born. As examples, the female swordtail and guppy will both give birth to anywhere from 20 to 100 live young after a gestation period of four to six weeks, and mollies will produce a brood of 20 to 60 live young after a gestation of six to 10 weeks.

Nutrition during Pregnancy

Other terms relating to pregnancy in fish relate to the differences in the mode and extent of support the female gives the developing offspring.

"Lecithotrophy" (yolk feeding) occurs when the mother provisions the oocyte with all the resources it needs prior to fertilization, so the egg is independent of the mother. Many members of the fish family Poeciliidae are considered to be lecithotrophic, however, research is increasingly showing that others are matrotrophic.

"Aplacentral viviparity" occurs when the female retains the embryos during the entire time of development but without any transfer of nutrients to the young. The yolk sac is the only source of nutrients for the developing embryo. There are at least two exceptions to this; some sharks gain nourishment by eating unfertilized eggs produced by the mother (oophagy or egg eating) or by eating their unborn siblings (intra-uterine cannibalism).

"Matrotrophy" (mother feeding) occurs when the embryo exhausts its yolk supply early in gestation and the mother provides additional nutrition. Post-fertilization transfer of nutrients has been reported in several species within the genera *Gambusia* and *Poecilia*, specifically, *G. affinis*, *G. clarkhubbsi*, *G. holbrooki*, *G. gaigei*, *G. geiseri*, *G. nobilis*, *P. formosa*, *P. latipinna*, and *P. mexicana*.

Viviparous fish have developed several ways of providing their offspring with nutrition. "Embryotrophic" or "histrotrophic" nutrition occurs by the production of nutritive fluid, uterine milk, by the uterine lining, which is absorbed directly by the developing embryo. "Hemotrophic" nutrition occurs through the passing of nutritive substances between blood vessels of the mother and embryo that are in close proximity, i.e. a placenta-like organ similar to that found in mammals.

Comparison between Species

There is considerable variation between species in the length of pregnancy. At least one group of fish has been named after its pregnancy characteristics. The surfperch, genus *Embiotoca*, is a saltwater fish with a gestation period of three to six months. This lengthy period of pregnancy gives the family its scientific name from the Greek "embios" meaning "persistent" and "tokos" meaning "birth".

The table below shows the gestation period and number of young born for some selected fish.

Species		Reproduction method	Gestation period (Days)	Number of young (Average)
Atlantic sharpnose shark	(*Rhizoprionodon terraenovae*)	Viviparous	300-330	4-6
Barbeled houndshark	(*Leptocharias smithii*)	Viviparous[a]	>120	7
Blackspot shark	(*Carcharhinus sealei*)	Viviparous[b]	270	1-2
Blue shark	(*Prionace glauca*)	Viviparous	270-366	4-135
Bonnethead shark	(*Sphyrna tiburo*)	Viviparous[c]		4-12
Bull shark	(*Carcharhinus leucas*)	Viviparous	366	4-10
Butterfly goodeid	(*Ameca splendens*)	Viviparous	55-60	6-30
Caribbean sharpnose shark	(*Rhizoprionodon porosus*)	Viviparous		2-6
Daggernose shark	(*Isogomphodon oxyrhynchus*)	Viviparous	366	2-8
Lemon shark	(*Negaprion brevirostris*)	Viviparous	366	18 (max)
Oceanic whitetip shark	(*Carcharhinus longimanus*)	Viviparous	366	1-15
Dwarf seahorse	(*Hippocampus zosterae*)	Viviparous	3-55	10
Sandbar shark	(*Carcharhinus plumbeus*)	Viviparous	366	8
Spadenose shark	(*Scoliodon laticaudus*)	Viviparous[d]	150-180	6-18
Viviparous eelpout	(*Zoarces viviparus*)	Viviparous[e]	180	30-400
Basking shark	(*Cetorhinus maximus*)	Ovoviviparous	>366	unknown[f]
Bat ray	(*Myliobatis californica*)	Ovoviviparous	270-366	2-10
Coelacanth	(g. *Latimeria*)	Ovoviviparous	>366	
Blue stingray	(*Dasyatis chrysonota*)	Ovoviviparous	270	1-5
Bluespotted stingray	(*Neotrygon kuhlii*)	Ovoviviparous	90-150	1-7
Carpet sharks	(f. Ginglymostomatidae)	Ovoviviparous	180	30-40
Knifetooth sawfish	(*Anoxypristis cuspidata*)	Ovoviviparous	150	6-23
Nurse shark	(*Ginglymostoma cirratum*),	Ovoviviparous	150	21-29
Sailfin molly	(*Poecilia latipinna*)	Ovoviviparous	21-28	10-140
Salmon shark	(*Lamna ditropis*)	Ovoviviparous	270	2-6
Sand tiger shark	(*Carcharias taurus*)	Ovoviviparous	270-366	2[g]
School shark	(*Galeorhinus galeus*)	Ovoviviparous	366	28-38
Shortfin mako shark	(*Isurus oxyrinchus*)	Ovoviviparous	450-540	4-18
Spotted eagle ray	(*Aetobatus narinari*)	Ovoviviparous	366	4
Tiger shark	(*Galeocerdo cuvier*)	Ovoviviparous	480	10-80
Tawny nurse shark	(*Nebrius ferrugineus*)	Aplacental viviparity		1-2

- Unlike any other shark, the yolk-sac placenta is globular or spherical.

- At first, the embryos are sustained by a yolk sac, but later a placenta develops.

- A bonnethead female produced a pup by parthenogenesis in 2001.

- The spadenose shark has the most advanced form of placental viviparity known in fish, as measured by the complexity of the placental connection and the difference in weight between the egg and the newborn young.

- The eelpout suckles its young embryos while still within their mother's body, making it the only fish species to suckle its offspring.

- Only one pregnant female is known to have been caught; she was carrying six unborn young.

- 1 per uterine horn

Poeciliopsis

Members of the genus *Poeciliopsis* (amongst others) show variable reproductive life history adaptations. *P. monacha* can be considered to be lecithotrophic because the female does not really provide any resources for her offspring after fertilization. *P. lucida* shows an intermediate level of matrotrophy, meaning that to a certain extent, the offspring's metabolism can actually affect the mother's metabolism, allowing for increased nutrient exchange. *P. prolifica* is considered to be highly matrotrophic, and almost all of the nutrients and materials needed for foetal development are supplied to the oocyte after it has been fertilized. This level of matrotrophy allows *Poeciliopsis* to carry several broods at different stages of development, a phenomenon known as superfetation.

P. elongata, *P. turneri* and *P. presidionis* form another clade which could be considered an outgroup to the *P. monacha*, *P.lucida*, and *P. prolifica* clade. These three species are very highly matrotrophic – so much so that in 1947, C. L. Turner described the follicular cells of *P. turneri* as "pseudo-placenta, pseudo-chorion, and pseudo-allantois".

Guppy

Guppies are highly prolific livebearers giving birth to between five and 30 fry, though under extreme circumstances, she may give birth to only one or two or over 100. The gestation period of a guppy is typically 21–30 days, but can vary considerably. The area where a pregnant guppy's abdomen meets the tail is sometimes called the "gravid patch", or "gravid spot". When pregnant, there is a slight discoloration that slowly darkens as the guppy progresses through pregnancy. The patch first has a yellowish tinge, then brown and become deep orange as the pregnancy develops. This patch is where the fertilized eggs are stored and grow. The darkening is actually the eyes of the developing baby guppies and the orange tinge is their jelly-like eggs.

Elasmobranchs

The majority of elasmobranchs are viviparous and show a wide range of strategies to provide their offspring with nourishment and respiratory requirements. Some sharks simply retain their young in the dilated posterior segment of the oviduct. In its simplest form, the uterus does not provide any additional nutrients to the embryos. However, other elasmobranchs develop secretory uterine villi that produce histotroph, a nutrient which supplements the yolk stores of the oocyte. Uterine secretions are perhaps most advanced in the stingrays. Following depletion of the yolk, the uterine lining hypertrophies into secretory appendages termed "trophonemata". The process by which the uterine secretions (also known as uterine milk or histotroph) are produced resembles that of breast milk in mammals. Furthermore, the milk is rich in protein and lipid. As the embryo grows, vascularisation of the trophonemata enlarges to form sinusoids that project out to the surface to form a functional respiratory membrane. In lamnoid sharks, following yolk use, the embryos develop teeth and eat eggs and siblings within the uterus. There is usually one fetus per uterus and it grows to enormous proportions of up to 1.3 m in length. In placental sharks, the yolk sac is not withdrawn to become incorporated into the abdominal wall. Rather, it lengthens to form an umbilical cord and the yolk sac becomes modified into a functional epitheliochorial placenta.

Male Pregnancy

Pregnant male seahorse

The male fishes of seahorses, pipefishes, weedy and leafy sea dragons (Syngnathidae) are unusual as the male, rather than the female, incubates the eggs before releasing live fry into the surrounding water. To achieve this, male seahorses protect eggs in a specialized brood pouch, male sea dragons attach their eggs to their legs, and male pipefish may do either.

When a female's eggs reach maturity, she squirts them from a chamber in her trunk via her ovipositor into his brood pouch or egg pouch, sometimes called a "marsupium". During a mammalian pregnancy, the placenta allows the female to nourish her progeny in the womb, and remove their waste products. If male pipefish and seahorses provide only a simple pouch for fish eggs to develop and hatch, it might not fully qualify as bona-fide pregnancy. However, current research suggests that in syngnathid species with well developed brood pouches, males do provide nutrients, osmoregulation and oxygenation to the embryos they carry.

Seahorse

When mating, the female seahorse deposits up to 1,500 (average of 100 to 1,000) eggs in the male's pouch, located on the ventral abdomen at the base of the tail. Male juveniles develop pouches when they are 5–7 months old. The male carries the eggs for 9 to 45 days until the seahorses emerge fully developed, but very small. The number born maybe as few as five for smaller species, or 2,500 for larger species. A male seahorse's body has large amounts of prolactin, the same hormone that governs milk production in pregnant mammals and although the male seahorse does not supply milk, his pouch provides oxygen as well as a controlled-environment.

When the fry are ready to be born, the male expels them with muscular contractions, sometimes while attaching himself to seaweed with his tail. Birth typically occurs during the night, and a female returning for the routine morning greeting finds her mate ready for the next batch of eggs.

Pipefish

The subcaudal pouch of the male black-striped pipefish

Pipefish brood their offspring either on distinct region of its body or in a brood pouch. Brood pouches vary significantly among different species of pipefish, but all contain a small opening through which the female's eggs can be deposited. The location of the brood pouch can be along the entire underside of the pipefish or just at the base of the tail, as with seahorses. Pipefish in the genus *Syngnathus* have a brood pouch with a ventral seam that can completely cover all of their eggs when sealed. In males without these pouches, eggs adhere to a strip of soft skin on the ventral surface of their bodies that does not contain any exterior covering – a type of "skin brooding".

At least two species of pipefish, *Syngnathus fuscus* and *Syngnathus floridae*, provide nutrients for their offspring.

The table below shows the gestation period and number of young born for some selected seahorses.

Species		Reproduction method	Gestation period (Days)	Number of young
Big-belly seahorse	(*Hippocampus abdominalis*)	Ovoviviparous	28	600-700
Lined seahorse	(*Hippocampus erectus*)	Ovoviviparous	20-21	650 (max)
Long-snouted seahorse	(*Hippocampus guttulatus*)	Ovoviviparous	21	581 (max)

References

- Romer, Alfred Sherwood; Parsons, Thomas S. (1977). The Vertebrate Body. Philadelphia, PA: Holt-Saunders International. pp. 385–386. ISBN 0-03-910284-X

- Chapman, D. D.; Firchau, B.; Shivji, M. S. (2008). "Parthenogenesis in a large-bodied requiem shark, the blacktip". Journal of Fish Biology. 73 (6): 1473–1477. doi:10.1111/j.1095-8649.2008.02018.x

- Lodé, T. (2012). "Oviparity or viviparity? That is the question…". Reproductive Biology. 12 (3): 259–264. doi:10.1016/j.repbio.2012.09.001. Retrieved November 4, 2014

- Fricke, Hans; Fricke, Simone (1977). "Monogamy and sex change by aggressive dominance in coral reef fish". Nature. 266 (5605): 830–832. Bibcode:1977Natur.266..830F. PMID 865603. doi:10.1038/266830a0

- Helfman G, Collette BB, Facey DH and Bowen BW (2009) The Diversity of Fishes: Biology, Evolution, and Ecology p. 35, Wiley-Blackwell. ISBN 978-1-4051-2494-2

- Pierce, S. J.; Pardo, S. A.; Bennett, M. B. (2009). "Reproduction of the blue-spotted maskray Neotrygon kuhlii (Myliobatoidei: Dasyatidae) in south-east Queensland, Australia". Journal of Fish Biology. 74 (6): 1291–308. PMID 20735632. doi:10.1111/j.1095-8649.2009.02202.x

- Theodore W. Pietsch. "Precocious sexual parasitism in the deep sea ceratioid anglerfish, Cryptopsaras couesi Gill". Nature. 256: 38–40. doi:10.1038/256038a0. Archived from the original on 28 August 2008. Retrieved 31 July 2008

- Robertson, D.R.; R.R. Warner. "Sexual patterns in the labroid fishes of the Western Caribbean II: the parrotfishes (Scaridae)". Smithsonian Contributions to Zoology. 255: 1–26. doi:10.5479/si.00810282.255

- Blaber, Stephen J. M. (2000) Tropical estuarine fishes: ecology, exploitation and conservation John Wiley and Sons, Page 153–156. ISBN 978-0-632-05655-2

- Seebacher, F.; Ward, A.J.W & Wilson, R.S. (2013). "Increased aggression during pregnancy comes at a higher metabolic cost". Journal Experimental Biology. 216: 771–776. doi:10.1242/jeb.079756. Retrieved November 2, 2014

- Moe, M. "The breeder's net: Science, biology, and terminology of fish reproduction: Reproductive modes and strategies-Part 1". Advanced Aquarist. Retrieved November 1, 2014

- Stölting, K.N. & Wilson, A.B. (2007). "Male pregnancy in seahorses and pipefish: beyond the mammalian model". BioEssays. 29 (9): 884–896. PMID 17691105. doi:10.1002/bies.20626

- Carrier, J.C.; Musick, J.A.; Heithaus, M.R., eds. (2012). Biology of Sharks and Their Relatives. CRC Press. pp. 296–301. ISBN 1439839247

- Hanel, R.; M. W. Westneat; C. Sturmbauer (December 2002). "Phylogenetic relationships, evolution of broodcare behavior, and geographic speciation in the Wrasse tribe Labrini". Journal of Molecular Evolution. 55 (6): 776–789. PMID 12486536. doi:10.1007/s00239-002-2373-6

- Jones, A.G. & Avise, J.C. (2003). "Male pregnancy". Current Biology. 13 (20): R791. PMID 14561416. doi:10.1016/j.cub.2003.09.045. Retrieved November 1, 2014

- Wourms, J.P. (1993). "Maximization of evolutionary trends for placental viviparity in the spadenose shark, Scoliodon laticaudus". Environmental Biology of Fishes. 38: 269–294. doi:10.1007/BF00842922

- Compagno, L.J.V. (2002). Sharks of the World: An Annotated and Illustrated Catalogue of Shark Species Known to Date (Volume 2). Rome: Food and Agriculture Organization. ISBN 92-5-104543-7

Permissions

Index

www.ingramcontent.com/pod-product-compliance
Lightning Source LLC
Chambersburg PA
CBHW080245230326
41458CB00097B/3423